WHISKEY

A CONNOISSEUR'S GUIDE

To Mum,
Thanks for always having faith in me.

THIS IS A CARLTON BOOK

This edition published in 1998

10 9 8 7 6 5 4 3 2 1

Copyright © Carlton Books Limited 1998

ISBN 1 85868 706 3

Project editor: Martin Corteel
Project art direction: Diane Spender
Picture research: Justin Downing
Production: Sarah Schuman
Design: Mary Ryan

Printed in Italy

WHISKEY

A CONNOISSEUR'S GUIDE

Dave Broom

CARLTON

CONTENTS

INTRODUCTION

It was in Tunisia where I fully appreciated the power of the word whisky. I had spent the first part of a holiday explaining that I was from "Ecosse". The standard response was a blank, uncomprehending stare. After a week of this I had come to the conclusion that Scotland was simply not known in North Africa, that I would have to grit my teeth and say that I was British. Then by chance, as yet another Tunisian tried to unravel where my mysterious homeland was I said, in exasperation, "whisky!" It was like a magical charm. Suddenly I was everyone's best friend. Ecosse wasn't known, but "Scotch", and "whisky" were. The next thing I knew it was four o'clock in the morning, I was covered in garlands of jasmine flowers, an empty bottle of Johnnie Walker Black Label on the table and another on the way.

That incident brought home how whisky has transcended boundaries, has become the world's favourite spirit, has become the currency and the lingua franca of a new nation. Just as you can discuss football with anyone in the world whether you can speak their language or not, so whisky connoisseurs can understand each other by dropping names like Lagavulin, Johnnie Walker, Power's, Wild Turkey and Jack Daniel's into the conversation.

It wasn't always this way. At the start of the 1970s malt whisky was basically unknown, as were the best products from Ireland and Kentucky. Blended Scotches were on the slide. Whisky, after a century of bestriding the world of spirits like a colossus, was looking distinctly shaky. These days there are more malt brands than ever before, as well as new improved blends. Bourbon has been revived, as has Irish whiskey. Why the turnaround?

The simple answer is that consumers the world over demanded flavour in their drinks. They were tired of being fobbed off with light tasting, image-driven, interchangeable brands. They wanted something which had history, authenticity and flavour – a drink that they could sink their teeth into. Malt whisky blazed the new trail and found a ready audience. The others soon followed. Suddenly whisky was no longer a commodity, but a host of highly individual drinks. The floodgates were open.

This book attempts to show the differences between the drinks which are satisfying this new audience. Where they come from, how they acquire their individuality, how you can taste them. All it takes to start on the journey to become a whisky connoisseur is one little sip. Get ready!

Dave Broom

Brighton, October 1998

THE STORY OF WHISKY

To FIND THE ORIGINS OF WHISKY, YOU HAVE TO DELVE AS DEEP AS YOU CAN IN THE REMNANTS OF THE CELTIC PAST, DIG UP THE FEW CLUES THAT LIE GLIMMERING THERE... AND THEN DRAW YOUR OWN CONCLUSIONS.

No one knows who the first whisky-maker was. In fact, no one knows who the first distiller was! What can be ascertained is that distillation started in either China or India, and spread to Egypt and Greece by the fourth century AD. With the exception of one reference in China, no one used the art of distillation to make spirits – medicine and perfumes were the norm.

The question is how the knowledge arrived in Ireland and Scotland, apparently bypassing the rest of Europe. It's possible that, if distillation sprang up in China and spread through India and the Caucasus, that the Celts – who were originally a tribe from this area – may have carried the art with them on their wanderings through Europe.

However it arrived there, distillation was certainly known by the monks of the Celtic Church in Ireland. They retained the secret during the Dark Ages and then took it – and the word of God – out into mainland Europe and Scotland, probably first distilling on Islay. There's no surprise that the first Scottish distiller to be mentioned by name was a monk Friar John Cor who, in 1494, was granted the right to make the equivalent of 400 bottles of whisky.

Ireland and Scotland are physically quite similar. Both have an abundance of pure water, plenty of peat to fire the stills and, most importantly, lands where barley could grow. The early distillers the world over have always made their spirit from the most widely available crop and in time, when whisky-making crept out from behind monastery walls, farmers leapt on it as a new way in which to use up surplus grain, feeding their wintering cattle and making a little money to help pay the rent.

Malted barley was the most commonly used grain in the Scottish Highlands and Islands, but oats and (importantly) unmalted barley were used in Ireland, while the

RIGHT

IN THE WORDS OF ROBERT BURNS: "FREEDOM AND WHISKY GANG THEGITHER"

distillers in the Scottish Lowlands used whatever cereal they could get their hands on. In Scotland, this division was highlighted after the 1745 Rebellion when the Highland way of life was banned, driving many crofter-distillers from Scotland across to America and Canada. At the same time great waves of Irish emigrants were forced off their land and ended up across the Atlantic. Whisky went with both groups.

Spirits were distilled in the new American colonies virtually from the word go – records show that rye was being distilled in Staten Island in 1640 – but it was this eighteenth-century wave of immigrants which established whiskey as North America's spirit. They immediately began making whiskey from the crops around them – rye in Pennsylvania and corn in Kentucky and Tennessee and around the Great Lakes.

Meanwhile, distillers in rural Ireland and the Scottish Highlands were bearing the brunt of punitive government measures which banned the use of small stills. Some distillers carried on, illegally, but the industry was being increasingly controlled by an alliance of government, rich businessmen and landowners. By

1823, when small distillers were encouraged to take out licences, whisky had become an industry. It was too late for Scottish malt and Irish whiskey though. The continuous still had been invented, and while Ireland fought a brave rearguard action against the new light whiskey it produced, malts became subsumed in the new blends that were being promoted world-wide by enthusiastic young Victorians.

Blended Scotch received its greatest boost when phylloxera wiped out brandy production in Europe. Then Ireland became independent (and lost sales to the British Empire) and America banned alcohol altogether! Scotch blends strode on. Even during the Second World War, when the Irish and American industries closed down, Scotland was producing small, but significant, amounts of whisky.

It has only been in the past 15 years that the picture has changed. Although blends still dominate, drinkers across the world are looking for high-quality products with flavour and heritage. Malt whisky has boomed, and so have Irish whiskey, and bourbon. Never has the connoisseur had so much choice.

BELOW

BEYOND THE REACH OF THE DONEGAL EXCISEMEN

CHRONOLOGY OF WHISKY

C. 500   Welsh bard Taliesen's "Mead Song" mentions distillation.

1174   Henry II of England records use of *aqua vitae* in Ireland.

1296   Sir Robert Savage, the Lord of Bushmills, gives *aqua vitae* to his troops.

1494   The first mention of whisky-making in Scotland.

1590   Whisky exports from Scotland begin.

1640   Rye whiskey distilled on Staten Island.

1644   The Scottish Parliament levies the first tax on whisky.

1717–18   First wave of Scots and Irish settlers arrive in North America.

1745   Start of the Highland Clearances, precipitating the next wave of Scots into North America.

1770–   Settlers arrive in Kentucky and Tennessee.

1780   John Jameson establishes distillery in Dublin's Bow Street.

1783   Evan Williams starts distilling in Louisville, Kentucky.

1784   The Wash Act in Britain restricts the sale of malt whisky to above the Highland Line.

1791–92   The Whiskey Rebellion in the United States.

1795   Jacob Beam founds his distillery.

1800   Two thousand distilleries in Ireland.

1820   John Walker opens his grocery shop in Kilmarnock.

1823   James (Jim) Crow begins working at the Oscar Pepper distillery, Kentucky.

        Wash Act introduces licences for distilleries in Scotland and Ireland.

1825   First mention of The Lincoln County Process (charcoal filtration) in Tennessee.

1827   Robert Stein invents a continuous still for whisky production.

1831   Aeneas Coffey improves on Stein's continuous still and patents it.

1837   William Teacher opens his first grocery store in Glasgow.

1845–49   The Great Famine drives more Irish to America and reduces the amount of grain available for distilleries.

1846   John Dewar starts up as wine and spirit merchant in Perth.

1853   The first Scotch blend, Usher's Old Vatted Glenlivet, is launched.

1867   Walker's Old Highland Whisky copyrighted and trademarked.

1887   Twenty-eight distilleries in Ireland.

1918–19   Prohibition in Canada.

1919   Prohibition in USA.

1933   Prohibition repealed in USA.

1941   US and Irish distilleries stop production.

1966   Jameson, Power and Cork Distillers merge to form Irish Distillers.

1973   Bushmills joins Irish Distillers.

1987   Cooley Distillers founded, breaking Irish Distillers' monopoly.

1998   UD and GrandMet (IDV) merge to form Diageo, the world's largest spirits firm.

ABOVE

HENRY II – BEATEN BY WHISKY-
FUELLED IRISH TROOPS

11

THE HEART
OF THE SPIRIT

ALL THE WORLD'S WHISKIES ARE MADE FROM GRAIN, WATER AND YEAST THEN AGED IN OAK CASKS. THE ART LIES IN THE WAY IN WHICH DISTILLERS HAVE TAKEN THIS SIMPLE FORMULA AND MADE SUCH A VAST ARRAY OF STYLES.

Different grains will be used in each country. These will be ground to a coarse flour and then have hot water poured on top of them, stripping the sugars from the flour. The sweetish liquid is drained off and pumped into large fermenters, where yeast is added and the mixture is left to ferment. After a couple of days the distiller is left with a strong, crude beer which is distilled.

Distillation can take place either in pot or column stills. In pot stills, distillation has to take place twice, sometimes three times, before the spirit is of sufficient strength and purity. It's during the second and third distillations where the stillman separates the heart of the spirit – containing the flavour compounds called congeners – from the volatile heads and heavy tails.

The shape of the still is of vital importance in creating the whisky's complex flavours, but ensuring that the distillation is done slowly is equally important. Too fast and the heads, heart and tails become jumbled up, resulting in a poor-quality spirit.

A column still works on the principle that if you put wash in at the top of the still and pump steam in from the bottom, the alcohol will be vaporized by the rising steam, carried back up and then condensed. This can be done in a single, or linked column.

Both types of spirit are then aged in oak casks, during which time the whisky moves in and out of the

ABOVE
THE STARTING POINT

OPPOSITE
KEEPING A WATCHFUL
EYE OPEN

OPPOSITE
STILL LIFE

wood, extracting flavours and colour from the oak and gaining in complexity.

This pattern is repeated day-in, day-out in distilleries across the world, each one with its own spin on proceedings. In this way each country has established its own point of difference and each distillery has created its own signature.

Malt

Scottish malt whisky is made exclusively from malted barley. The process starts with the barley being germinated, triggering enzymes which convert its starch into soluble sugar. The germination is then stopped by drying the grain (usually over peat) in a kiln. The more peat you use, the peatier the malt.

The malt is either double- or triple-distilled in pot stills and the new spirit is then placed in used oak casks. The type of wood and what the cask previously held will impart different flavours. Ex-sherry barrels give a sweet richness and dark colour to the whisky, while ex-bourbon barrels give toasty vanillin characters.

BELOW
MAKING THE CUT

Grain Scotch

Grain whisky is usually made from maize which has had a little malted barley added to it before being distilled in linked column stills. This gives a high-strength spirit that's delicately flavoured but not neutral.

Irish Whiskey

Irish whiskey is subtly different from Scotch, not only using a percentage of unmalted as well as malted barley, but approaching distillation differently.

At Irish Distillers' Midleton distillery each brand has its own recipe of malted and unmalted barley, which is then triple-distilled in linked pot and column stills. Bushmills uses only triple pot-still distillation, but also imports some grain whiskey from Midleton. Neither of these distillers uses peat, but newcomer Cooley has revived the technique – as well as using double-distillation. Most Irish whiskey bourbon barrels, but sherry and rum ca

Bourbon

A bourbon must be made up of a min cent corn, the rest of the mash bill be small grains – rye, malted barley and v tiller has its own recipe.

The next main difference in produc tion of backset (the acidic spent wash) t to allow a clean, infection-free fern exception all American whiskey is first gle column still and then redistilled i known as a doubler.

In maturation, only new, charred

barrels are used, placed in massive multi-storey ware-houses. Because of the hot summers, the whiskey extracts the wood flavourings at a different rate on each storey. To ensure a consistently flavoured product, distillers either rotate the barrels to get an even maturation, or blend between the different levels.

Tennessee whiskey can be made from a minimum of 51 per cent of any grain and the new spirit has to be filtered through maple charcoal. This leaches out impurities and also gives the new spirit a sweet flavour.

Canadian whisky

Canadian whisky is distilled in column stills, but distillers approach the process slightly differently as they distil different grains (rye, corn, wheat, barley) either individually or together. These are distilled to different strengths, and matured in a wide range of woods for differing lengths of time. Canadian whisky's secret lies in the blending of these different components.

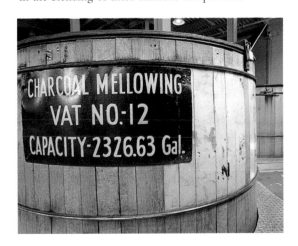

LEFT

THE DETAIL THAT MAKES
TENNESSEE SPECIAL

15

SCOTTISH WHISKY

TAKE ONE SMALL COUNTRY SITUATED IN THE FAR NORTH-WEST CORNER OF EUROPE. STIR IN THE WEATHER, GEOLOGY AND AGRICULTURE. THEN BLEND IN UNKNOWN QUANTITIES OF HUMAN DESIRE, ABILITY AND SHEER BLOODY-MINDEDNESS, AND WAIT. THE RESULT IS SCOTCH.

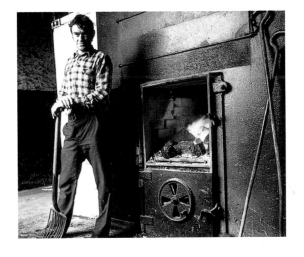

LEFT
STOKING THE FIRE ON ISLAY

Scotland is blessed with a climate that regularly dumps plenty of water onto its hills and glens. It may make you bemoan your luck if you are stuck in the Highlands on summer holiday, being eaten alive by midges, unable to see the tops of the mountains, but the plentiful rain is the first element in making high-quality whisky.

Having percolated deep underground, this water filters back to the surface through rocks that are rich in mineral deposits. Some of the resulting burns flow over peat, the dark fuel that lends its heavy smell to many malts.

This country is well-suited to growing barley, the next element in making malt whisky. The raw materials are on hand: the final element is man.

Scotland is home to a race of people who have always loved their drink. It claims to be the first coun-

try to import wines from Bordeaux, and its beers have long been held in international renown, so there's no great surprise that when the secret of distillation was discovered by the people of the Highlands, they applied themselves to learning this strange alchemy.

Other countries make whisky – Ireland was probably the first, America and Canada have long made excellent examples – but Scotland remains its spiritual home. Why? Chance certainly plays its part, but so does the arrival of blended whisky at the very time when Scotland was the engine-room of the British Empire. If Highland distillers perfected malt, then it was Victorian entrepreneurs who conquered the world with blends.

Today, connoisseurs can enjoy the greatest selection of Scotch whiskies ever. Here are some of them.

OPPOSITE
THE AROMAS OF THE
SCOTTISH WILDERNESS ARE
CAPTURED IN THE GLASS

THE LOWLANDS

THE LOWLANDS AREN'T IMMEDIATELY ASSOCIATED WITH MALT. SURELY WHISKY COMES FROM THE HIGHLANDS? AREN'T THE LOWLANDS SCOTLAND'S OVERPOPULATED, INDUSTRIALISED HEARTLAND? WELL, MOST OF THE HEAVY INDUSTRY HAS LONG GONE AND THE URBAN SPRAWL HAS ALWAYS BEEN SURROUNDED BY GENTLE HILLS, MOORS AND FARMLAND.

ABOVE

AS FRESH AS A SUMMER'S DAY

RIGHT

AUCHENTOSHAN PRODUCES A CLASSIC LOWLAND STYLE

That said, the bulk of the spirits produced in this part of the country is grain whisky for blends, but there are other factors which make Lowland malt subtly different from its northern neighbours.

The distinctive Lowland style

There has always been a wide variety of cereal crops grown in these fertile lands, and in the past many distillers used a mix of different grains rather than just barley. There wasn't much peat here to dry the barley, but there was coal, and distillers also often triple-distilled.

The Lowland style therefore emerged as one that was soft, dry and relatively light in character.

This isn't being dismissive. Good malt whisky has always been produced here, often in distilleries within grain plants, such as Grant's Ladyburn, or Inverleven and Lomond from the innards of Allied's Dumbarton plant.

These days, though, there's the merest smattering of malt distilleries. Auchentoshan, situated next to the Clyde on the outskirts of Glasgow, still produces a fine triple-distilled malt. UDV's Glenkinchie is one of the firm's Classic Malts and is a pleasant, hay-like dram. In fact, Glenkinchie was the lucky one out of UDV's Lowland plants. A slump in demand meant it had to close Bladnoch, Scotland's most southerly distillery, and Rosebank, which produced the finest of all Lowland malts. The good news is that

ABOVE

SPRINGBANK: HOME OF SOME
OF THE GREATEST MALTS OF
ALL

both look likely to re-open (in a limited way) soon, enabling us to enjoy more examples of a style that's as charming as this part of the country.

The Campbeltown saga

Technically speaking, Campbeltown is a separate region, although these days it has only one distillery in operation. It wasn't always like this. At one time Campbeltown was Scotland's whisky capital. At the end of the last century there were 25 distilleries cluttering up this small fishing port at the foot of Kintyre. Campbeltown's speedy decline was down to a combination of greed and Al Capone and his buddies. The town had supplied America with whisky for years, and when Prohibition struck the orders increased. Trouble was, as production was upped, so quality fell. The bootleggers also realized they could get more money for any old hooch by slapping Campbeltown on the label. The town's reputation plummeted, and by 1930 there were only three distilleries left.

Somehow Springbank, the only Campbeltown distillery still open, has managed to remain a magnificent dram. It still malts its own barley, uses coal-fired stills, a form of triple-distillation, and even bottles, unfiltered, on the premises. If only Campbeltown Loch were full of this stuff! Truly magnificent.

LOWLANDS AND CAMPBELTOWN

THE WHISKIES FROM THE LOWLANDS ARE A PERFECT INTRODUCTION TO MALT. PEATING LEVELS ARE KEPT TO A MINIMUM AND TRIPLE DISTILLATION IS COMMON – TECHNIQUES WHICH PRODUCE WHISKIES THAT HAVE A DELICATE, CHARMING SURFACE. THAT DOESN'T MEAN THEY HAVE MISSED OUT ON COMPLEXITY THOUGH. CAMPBELTOWN SHOULD BE CONSIDERED A REGION IN ITS OWN RIGHT AND ONE WHICH FUSES ALL OF MALT WHISKY'S GREATEST CHARACTERISTICS IN THE GLASS.

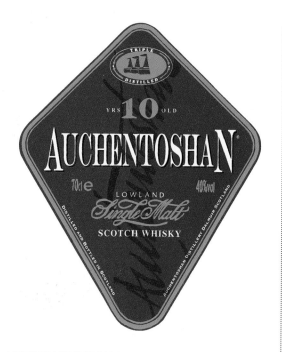

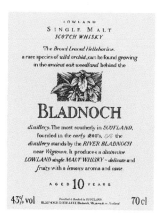

AUCHENTOSHAN

10-year-old (40% ABV)

Turfy, light and slightly sweet nose without water. Fresh, mealy and refreshing. A crisply attractive malt that's as fresh as a daisy with a long, clean finish.

AUCHENTOSHAN

21-year-old (43% ABV)

A nose of freshly cut turf. Good, nutty fruit with toasty aromas. Sweet, with a rich weight that rolls around the tongue before a rich, sweet finish. A revelation.

BLADNOCH

10-year-old (40% ABV)

Fresh, almost minty, nose with a hint of caramelized orange and hay. A delicate, slightly sweet start, before drying out a little mid-palate. Decent weight.

GLENKINCHIE

10-year-old (40% ABV)

A typical Lowland nose: grassy, crisp and invigorating. A touch of lemon on the palate intensifies the overall impression of freshness.

GLENKINCHIE

1986 Amontillado Finish (43% ABV)

Some delicate walnut aromas mingling with malt loaf and the signature grassiness. Rounder and sweeter than the 10-year-old with fruity tingle on the finish.

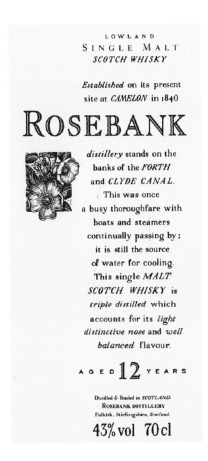

ROSEBANK

12-year-old (43% ABV)

Tremendous, complex nose with green grass, apple and an undercurrent of hay-accented fruit. Wonderfully balanced, with some smoke on the palate before a huge lift of sweet fruit that lasts forever. A classic.

SPRINGBANK

12-year-old (46% ABV)

An astounding, powerful, perfumed nose that is delicate and lightly smoky with positive, malty, lightly sherried notes. A sweet, hugely ripe start before dried spices and smoke peek out.

SPRINGBANK

1966 Local Barley (54.4% ABV)

Intense, powerful nose. This is a crunchy, mouth-watering mix of smooth, elegant wood, butter, lemon peel and peaches. Layer upon layer of flavours that grow in the mouth and stay there for eons.

ISLAY

T HIS, THE MOST SOUTHERLY OF THE SCOTTISH ISLANDS, IS HOME TO THE COUNTRY'S MOST DISTINCTIVE GROUP OF MALTS. IF FINDING A REGIONAL STYLE FOR PARTS OF THE MAINLAND SOMETIMES INVOLVES MAKING RATHER CREATIVE CONNECTIONS, THEN THERE'S NO DOUBT THAT THE MAJORITY OF ISLAY'S MIGHTY BROOD ARE FROM THE SAME FAMILY.

Whisky's cradle

Opinion is moving toward the conclusion that this friendly island could have been the first part of Scotland to make whisky. The southern tip of Islay is only 30 miles from the north of Ireland, and in the fifth century it was the site for the first landing of Christian missionaries from Ireland. Given it is held that these monks held the knowledge of distilling, there's every possibility that once they had established their monastic strongholds on the island they continued with their experiments in making medicinal spirits.

In the confusing world of early Scotland, Islay was part of a separate Norse kingdom. Even after the Vikings had relinquished control it was the seat of power for the Lords of the Isles who, though they swore fealty to the Scottish crown, ruled the Western

Isles as their own domain.

Much has been made of an apparent visit by King James IV to the island in the 1490s, just before he gave permission for Friar John Cor to make whisky. Might he have picked up the secret on Islay? It seems unlikely. Firstly, the King never visited Islay; in fact in the 1490s he was in the process of stripping the last Lord of the Isles of his title and installing his own man. He might have tasted whisky, but as far as being taught the secrets...?

Also, Friar Cor was making a huge amount of whisky, implying that the secret had long been known on the mainland. Islay may be the birthplace, but it is highly unlikely that it managed jealously to guard the secret for centuries.

Whatever the origins of distilling on Islay, the fact remains that it is particularly blessed for whisky making, one reason for it being able to still have seven distilleries when every other one of the Western Isle islands has, at most, one.

A benevolent climate

Islay is low-lying, fertile, the recipient of copious quantities of rain from the westerlies which strafe across it, but it also enjoys plenty of sunshine. Crofters on other

islands may have had to struggle with raising barley, but on Islay there was little problem. To make matters even better for the whisky-maker, half of the island is smothered in peat. Until recently, Islay was self-sufficient in whisky-making ingredients. During the smuggling era – though it's due west of Glasgow, Islay is in the Highlands – stills could be hidden in its shoreline caves and the whisky quickly transported out. When legal distilling arrived, the old farmer distillers banded together (as at Lagavulin) and continued to make their whisky on the coast. This meant that barley could be imported easily (when commercial distilling started, Islay could no longer provide sufficient barley to supply the distilleries) and barrels of whisky could be taken off the island by boat. Bunnahabhain, the remotest distillery on the island, got all of its supplies by boat until the 1970s.

Another important reason for the high survival rate of Islay's distilleries was the fact that the island was less badly affected by the Clearances. While the population of neighbouring Jura was decimated to make way for sheep and estates, Islay was reorganized, with crofts being absorbed into larger farms. Though people were forced to leave their land, virtually all found employment elsewhere on the island. The island therefore had

a relatively large, stable population – and one that appears to have had a fearsome thirst!

As the whisky industry began to realign itself as a mighty commercial machine, many of Islay's farm distilleries – like the one at Bridgend which now houses the Post Office and general store – fell by the wayside. That said, of the distilleries that Alfred Barnard visited in the 1880s only one, Lochindaal, no longer exists. Two, Bruichladdich and Port Ellen, are mothballed but there's every chance that the former will reopen soon, while

BELOW

THE MASSIVE BULK OF
CAOL ILA

Port Ellen now supplies virtually all of the island's malted barley. That's an amazing survival rate when you compare what happened in Campbeltown only 30 miles away. The reason is down to the singular flavour and high quality of Islay's malts.

The island falls quite neatly into two halves both geologically and stylistically. The younger southern part – home to Ardbeg, Lagavulin, Laphroaig and Port Ellen – contains the largest deposits of peat and, surprise surprise, the peatiest malts. The older northern half – home to Caol Ila, Bunnahabhain and Bruichladdich – is rockier and the style, generally speaking, is firmer and less heavily peated. Bowmore, the island's capital, sits halfway between – its water source is from the old north, and its peaty character nods to the south.

The influence of water

Water is a particularly important element in the flavour of Islay's malts. The southern group all take their water from burns which rise and flow over the peat bogs. The water that comes out of taps here is brown and with the faintest touch of a peaty, mossy flavour. In the north the water comes from springs or burns which have had little or no contact with peat.

Needless to say, the actual peating level given to the malted barley also makes a huge difference. Nowhere is malt given such a lengthy time with peat. Ardbeg's peating level is an astounding 50–56ppm, Laphroaig, Lagavulin and Caol Ila are around 35ppm, Bowmore is 20ppm while Bruichladdich and Bunnahabhain are at the industry average of 1–2ppm.

Significantly, all the distilleries use Islay peat either malted to their specifications at Port Ellen or, in Bowmore's and Laphroaig's case, malted in traditional floor maltings. You may think all peat is the same, but you'd be wrong. On the mainland it's made up of wood, moss and heather, but Islay's peat has layers of carbonized seaweed intertwined with the moss and heather. The wind also means that trees are an endangered species so wood-fired stills were never an option.

Because the island is regularly lashed with sea spray, the peat has a slight iodine, salty tang.

Stir all these components together and add in the fact that all the distilleries age their whisky in warehouses on the seashore – again increasing the interplay between whisky and salt-laden air – and you can begin to see why Islay remains so special.

The distiller's art

That's not to say that its malts are all identical. Far from it. This is where the art of the distiller comes into play. Take the threesome from the whitewashed citadels on the bleak, eerie south coast. Peaty to the core, each shows a subtle variation on the theme. Laphroaig is the oiliest and heaviest – filled with tarred ropes, seaweed, oilskins and not so much a peat reek as a damp fire smouldering in the glass. The standard 10-year-old is good, but the cask-strength is magnificent – like Talisker, this is whisky whose flavours need a high-strength foundation to express themselves fully.

Lagavulin is equally peaty and muscular, but has greater elegance, partly down to the use of sherry wood (Laphroaig uses only ex-Bourbon barrels); a deep and contemplative malt. Ardbeg, which you'd expect to be brutally, almost one-dimensionally peaty is instead superbly complex. The distillery was saved from an ignominious fate in 1997 when Glenmorangie snapped it up for a bargain price. Its future as a producer of magnificent malts looks assured. Port Ellen, which sadly looks unlikely ever to reopen, produced the most uncompromising of all Islay malts – one exclusively for the peat freak.

Variety in the north

The situation is more varied in the north of the island. Caol Ila shares the same peating regime as Lagavulin, but its malt is a considerably different beast. For years this massive plant was used by owner UDV to provide all manner of shadings of whisky – from hugely peaty to virtually unpeated. This was good for the blenders as it meant that they could use fewer components in a blend, but was pretty confusing for the malt aficionado. You were never quite sure what you'd be getting – one

bottling would smell of fish heads on the beach, the next would be rather thin and greasy. These days it seems to be on a more even keel, providing a single malt that shows good phenolic character, but allied to the firm style you get from the northern part of Islay.

Further up the same north-east coast is Bunnahabhain. Here, a combination of pure water, low peating and tall stills gives a malt which is miles away from what people consider to be a true Islay character. But while the reek may not be there, the malt has a delicious seashore freshness, a quality it shares with Bruichladdich. This marvellous old distillery, complete with ancient cast-iron mash tub, produced the malt that is most widely drunk on the island, and you can see why. There's some of the smoke that hangs in the Islay air on the nose, even a whiff of sea-shells drying in the sun at the high-water mark. It's not peaty, but somehow it could only come from this island.

Bruichladdich sits on the shores of Lochindaal, directly opposite the island's capital, Bowmore. This town is the axis around which the rest of the island spins, and its malt is a superb fusion of all of Islay's unique characteristics. A complex mix of woods, good peating levels and high-quality distilling, it's difficult to surpass.

Islay, like all of the islands, remains a place apart, an island with its own rules, its own character, its own way of approaching life. The concentration of distilleries has given whisky a central part in the island's economy – the whole community depends on it, not just the men who work in the distillery, but their wives and daughters who guide tourists round. It provides employment for local transport firms, while the farmers rely on the spent grains for cattle feed. Bowmore has even converted one of its old warehouses into the community swimming pool, whose water is heated by waste heat from the distillery. You'll not find anywhere else in Scotland where whisky plays such a central role.

ISLAY

THE CLASSIC MALTS

ISLAY REMAINS HOME TO SCOTLAND'S GREATEST COLLECTION OF PEATED MALTS. THE ISLAND'S VERY SOUL SEEMS TO BE ENCAPSULATED IN THE GLASS – THE SEA-LADEN AIR, THE REEK OF THE LOCAL PEAT, THE COMBINATION OF RUGGEDLY ROBUST FLAVOURS AND GENTLE FRUIT. THAT SAID, THERE ARE TWO WHICH HAVE LITTLE PEAT – THE BEST FOR THE ISLAY VIRGIN, WHILE THE REST, THOUGH RICH AND SMOKY HAVE A GRACEFUL POWER. ONCE TRIED, THESE ARE MALTS THAT ARE NEVER FORGOTTEN.

ARDBEG

17-year-old (40% ABV)

Orange and lemon marmalade nose. Chewy, with rich, integrated peatiness and balanced wood. Rich, with vanilla mingling with the fragrant smoke. Opulently flavoured, laced with lapsang souchong tea and ripe fruit.

ARDBEG

1978 (43% ABV)

Big, grumbly unreduced nose before the peat reek gracefully unfolds. A light touch of peat oil on the palate along with rich, smoky fruits and toffee, heather and salt. Glorious.

BOWMORE

17-year-old (43% ABV)

Intense, elegant nose that mingles peat smoke, Jaffa cakes and a malty lift. On the palate are cigar box, peat smoke and chocolate. Elegant, soft and very long.

BOWMORE

Darkest (43% ABV)

Glowing colour. Subtle and rounded nose of walnuts, chocolate and smoky fruit. Very well balanced.

BRUICHLADDICH

10-year-old (40% ABV)

Rounded, slightly salty nose with a light smokiness. With water, beachy aromas appear with a little bergamot. On the palate some smoke and then attractive, almondy fruit. Charming.

BUNNAHABHAIN

12-year-old (40% ABV)

Toffee-ish nose with a manzanilla-like salty tang. A bracing start, well rounded and gentle on the palate with a sudden ping of ginger on the finish.

CAOL ILA

15-year-old (43% ABV)

A mix of medicinal, rooty notes with lanolin and soft peatiness. Full on the palate with tobacco leaf and peat smoke. Crisp finish, with a touch of salt.

LAGAVULIN

16-year-old (43% ABV)

A great waft of fragrant smoke with some elegant woodiness, nutty fruit and tanned leather. Rich and plummy start before the peat surges back on the finish.

LAPHROAIG

10-year-old Cask Strength (57% ABV)

Hugely peaty nose with some fishy oiliness, a rich and powerful mix of seaweed and ozone. Very fresh malt on the palate, powerful and weighty. A classy dram.

LAPHROAIG

10-year-old (40% ABV)

Peat-sodden nose, medicinal with a light biscuity crunch. Great banks of peat, seaweed and trawler decks waft out. Oily, rich, salty, with some coconut on the finish.

WESTERN ISLES
AND ORKNEY

T HE ISLANDS THAT LIE SCATTERED IN THE SEA OFF SCOTLAND'S WESTERN AND NORTHERN COASTS ARE HOME TO SOME OF THE COUNTRY'S MOST INDIVIDUAL WHISKIES. THE ISLANDS BORE MUCH OF THE BRUNT OF THE CLEARANCES, WHICH STARTED AFTER THE FAILURE OF THE 1745 REBELLION AND REACHED THEIR PEAK IN THE MIDDLE OF THE NINETEENTH CENTURY. PEOPLE WERE DRIVEN OFF THE LAND AND EITHER FORCED TO LEARN NEW TRADES IN CUSTOM-BUILT SETTLEMENTS OR WERE HERDED ON TO BOATS SAILING TO AMERICA, CANADA AND AUSTRALIA.

It's easy to think that the islands were always the bleak, lonely, beautiful places they are today. But only 100 years ago they were filled with small communities, tending their crops and animals and making whisky. The production may not have been huge, but it helped to pay the rent. At one time whisky was made on Lewis, Pabbay and Tiree, while Skye alone could boast seven distilleries. Now, outside Islay, only five island distilleries remain.

The island distilleries remain places of pilgrimage. You don't chance upon them as you can on the mainland; you make a conscious decision to take the ferry and the often long drive to reach your destination. It's well worth the effort.

Jura

The largest building in the only settlement on the island, Jura's distillery has clung on while the

population of the island has declined. If it's isolation you want, head here.

Given its proximity to Islay, you might expect Jura's malt to be a big, peat-filled bruiser – the island's rugged landscape would imply that this should be a great hairy malt to sustain you over the last few miles of a hike. Instead, it's a little charmer, with only a wisp of peat and the lightest of salty tangs on the finish. The 10-year-old is an easy-drinking malt, pretty and commercial. Sadly its owner hasn't yet seen fit to release some of the remarkable older stuff that lurks in the warehouses. Here, given age and a bit of sherry wood, Jura comes into its own. There again, everyone forgets about Jura.

Mull

It's much the same with Mull. Although it's the second largest island in the Inner Hebrides, hardly anyone bothers to spend time here, other than using it as a stopping-off point for Iona, which lies off its western coast. Given it's likely that it was the monks of the early Celtic church who brought the art of distillation to Scotland, it's entirely possible that Mull was an early centre of distilling. By the late eighteenth century, up to a third of the island's barley harvest was turned into whisky and now-abandoned settlements, self-sufficient in barley, were important centres for whisky-making. By the time Barnard arrived, the sole distillery left was importing its grain from the mainland.

That distillery has had a chequered career. It's been closed for long periods, changed its name with baffling frequency from Tobermory to Ledaig (and back again) and had a series of absentee owners. Thankfully, it has been saved by Burn Stewart and now two malts – unpeated Tobermory and peated Ledaig are on offer. There are some good examples from independent bottlers – but given the ups and downs of Tobermory's recent past there are some shockers as well. Tread carefully.

OPPOSITE

THE WILD BEAUTY OF JURA

ABOVE

BACK ON THE RIGHT TRACK

Skye

This island, the largest in the Hebrides, has a humbling effect on the visitor. It doesn't throw its arms open to you, but forces you to love it on its own terms. The mountains, rarely free from straggling clouds, loom around you, making you feel very insignificant indeed. It's uncompromising land — and there's little surprise that Skye's only malt, Talisker, shares some of these qualities.

It's a peaty dram, but in a different way from that of Islay. The water flows over peat, and the barley is peated, but probably only to around 25ppm, about half that of Lagavulin. Why then does Talisker remain one of Scotland's most uncompromising malts? One reason may be the shape of the wash stills. They have high necks, a long lye pipe with a U-shaped kink in it and a purifier pipe that takes the heavier alcohols back into the still to be redistilled. This strange configuration increases reflux and also helps to give a long, slow condensation — as do the worm tubs situated outside the distillery.

You'll find it a peaty, smoky dram with touches of burnt heather root and black pepper, but also sweet fruit. There's some ozone freshness on the nose, but no iodine. The standard 10-year-old is matured in refill casks: there's no overt Bourbon or sherry wood here. It isn't fazed by high alcohol, in fact it needs it. It's an enigmatic dram, as craggily individual as its island birthplace.

Orkney

If Skye is the most elemental of Scotland's islands, then Orkney is the most other-worldly. The feeling on these islands is one of timelessness; it's a part of Scotland where the old ways have been absorbed and warped into the regular fabric of life.

Inevitably the church plays a central role in Orcadian life — and in the birth of its finest malt, Highland Park. During the smuggling era, the distillery was operated by one Magnus

Eunsen, a preacher who used to hide casks of whisky from the excise men beneath the pulpit.

Its top distillery, Highland Park, still malts some of its own barley using Orkney's heathery peat — an aroma which is carried through to the glass. Bundles of dried flowering heather are still added to the peat to impart an aroma to the malt. There's a judicious amount of (Spanish) sherry wood used, seen most noticeably in the 18- and 25-year-old versions. But even at 12 years of age, Highland Park is a superbly rounded malt. There's honeyed sweetness, smokiness, a little heather and a long, dry finish. It's a malt that can be drunk at any time of the day.

Orkney's other distillery, Scapa, has always been in the shadow of its famous neighbour on the hill. For long used as a filling station for blended whisky, its single malt is light and pretty, but not hugely memorable.

At the end of the day this is a disparate grouping of malts, whose only real connection is that they all come from islands. Yet it's not too fanciful to believe that the best of them have managed to weave the special qualities of their place of birth into the whisky.

ABOVE

BEHIND THE CALM EXTERIOR IS A BEAST OF A MALT

LEFT

AN ELEMENTAL WHISKY

OPPOSITE

ANOTHER CASK OF ORKNEY GOLD

WESTERN ISLES AND ORKNEY

THE CLASSIC MALTS

Every one of Scotland's many islands has its own distinctive feel. The same can be said for the relatively few malts which hail from this "region". The two camps – peated (traditional) and unpeated (modern) – give the island-hopping connoisseur plenty of scope for exploration. Here you'll encounter malts which seethe with almost volcanic power, others which are like a stroll along the sea-shore and on Orkney, a distillery which makes one of the most gorgeous malts of all.

HIGHLAND PARK

12-year-old (40% ABV)

Softly succulent nose of heather honey, ripe orangey fruit, hazelnuts, smoke and a light medicinal air. Rich, complex palate with heather, silky, fudgy fruit and a smoky finish.

HIGHLAND PARK

18-year-old (43% ABV)

An elegant, concentrated, rich nose with positive peat smoke. Gorgeous, supple palate with orange peel, sweet honeyed tones and a fragrant herbal lift. Superbly balanced with a long, smoky finish.

ISLE OF JURA

10-year-old (40% ABV)

Tight and a little woody on the nose, but clean with a nodule of sweetness on the palate. Fresh, decently balanced and a tingle of salt on the finish.

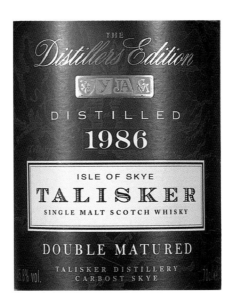

SCAPA

12-year-old (40% ABV)

A pleasant mix of sweet fruit and lightly toasted wood on the nose. Dry and clean with a tickle of peat smoke. A firm, nutty palate with good drive.

TALISKER

1986 Amoroso Finish (45.8% ABV)

Rich and smoky, but the elemental force of the 10-year-old is toned down. Rich sweetness, with dark, raisined fruit and peat reek.

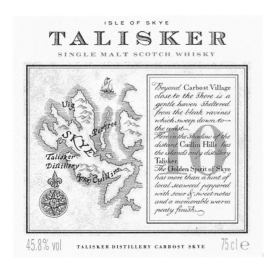

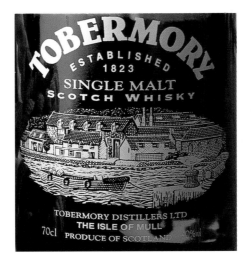

TALISKER

10-year-old (45.8% ABV)

A powerful, pungent nose with massive peatiness, charred heather, ozone and rich fruit. It explodes on to the palate, balancing sweet fruit with spicy, salty flavours before a black pepper kick on the finish.

TOBERMORY

(40% ABV)

A nose reminiscent of damp earth and nuts that carries through on to the palate, along with positive fruitiness and a little honey. Pretty simple, with a clean, firm finish.

SPEYSIDE

SPEYSIDE CONTAINS THE GREATEST CONCEN-
TRATION OF MALT WHISKY DISTILLERIES –
AND MANY OF THE BEST. ROUGHLY TRIAN-
GULAR IN SHAPE, IT UNFOLDS TO THE
NORTH ALONG THE WATERSHED OF THE RIVER SPEY,
STARTING AT THE EDGE OF THE GRAMPIAN MOUNTAINS
AND RUNNING DOWN TO THE FLAT, FERTILE COASTAL
PLAIN. THE ISOLATED, WILD AND BEAUTIFUL SOUTH-
ERN PART IS AS REMOTE AS ANY OTHER PART OF THE
HIGHLANDS, WHILE THE BULK OF THE DISTILLERIES IN
THE CENTRE AND NORTH ARE CLUSTERED AROUND
SOLID, RESERVED GREY STONE TOWNS. THE WHOLE
AREA IS RIBBONED BY FAST-RUNNING RIVERS, BURNS
AND SPRINGS THAT FLOW FROM THE PEAT-CAPPED
GRANITE MOUNTAINS. ALL IN ALL, IT'S PERFECT
WHISKY-MAKING COUNTRY.

BELOW

AULTMORE: BELOVED
BY BLENDERS – AND
CONNOISSEURS

Distilleries at every turn

The southern part of Speyside contains a smattering of distilleries, like Tomintoul and Braeval, before you reach The Glenlivet, the distillery that kick-started the region as one famed for high-quality malt. If you then follow the river Avon to its meeting with the Spey, you'll pass by neat and tidy Cragganmore and, further on, Glenfarclas. Then, distilleries begin to come thick and fast, every small road and glen bringing you face to face with well-known names like Tamdhu, Knockando and Cardhu. Heading into Dufftown, you'll pass by Dailuaine, Benrinnes and the great Aberlour – and up on the hill on the other side of the river catch a glimpse of the manor house that serves as Macallan's head office.

You can either then follow the course of the Dullan Water into Dufftown or travel north to Rothes. In Dufftown you can take your pick from seven distilleries – three of which, Glenfiddich, Balvenie and Kininvie are on one site (which also includes a cooperage). Less obvious is Mortlach, the distillery that established Dufftown as Speyside's capital.

Right enough, the good folk of Rothes would say that their town has a rightful claim to that title, as five distilleries are tucked away behind its rather dour facade (there's something very Scottish about that). Only the imposing Victorian buildings and elegant gardens of Glen Grant indicate that this is a whisky town. You'll have to search to find Caperdonich – not that it's worth the search: better to winkle out the underrated Glen Rothes and, hidden away up a tiny valley, Speyburn.

The last main whisky town is Keith, which lies roughly 10 miles to the east of Rothes. By now the heather-clad moors and hills have given way to fertile agri-

cultural land. To get to Keith you pass Auchroisk, which looks more like a modern hotel than a distillery, before arriving at the first of Keith's distilleries, the wonderful Aultmore, a distillery whose top-notch whisky is so loved by blenders that very little is allowed to slip out as a single malt. Although Strathmill and Glen Keith make pleasant enough drams, Keith's glory is the ludicrously picturesque Strathisla which produces one of Speyside's most charming and complex malts.

From there the long, flat Elgin road takes you north-west towards the coast, passing by two other overlooked classics, Longmorn and Linkwood. While Elgin has only one distillery, Glen Moray, the forgotten member of the Glenmorangie group, it is home to whisky's greatest retailer, Gordon & MacPhail, a Highland grocer and whisky specialist whose shelves groan under the weight of a mind-boggling range of malts.

Fame and fortune

That quick tour demonstrates what a concentration of top-notch distilleries is in Speyside. So why has this region become so famous? Whisky has flowed from

ABOVE

GLENFIDDICH'S SISTER – AND A BETTER MALT

ABOVE

JUST RESTING ... A GLENLIVET
WAREHOUSE

Speyside's stills for centuries, although no matter how hard you search, you won't discover when it was first made here. It's entirely possible that the large monasteries that were established in and around the area would have contained stills, but there's no hard evidence of this. Alan Winchester, the distillery manager at Aberlour and a man who combines Scottish pragmatism with the heart of a poet, likes to believe that the Celtic monk, St Drostan – who, handily enough, is buried under Abelour's still house – may have been the man who brought whisky to this part of Scotland. It's as good a theory as any other.

By the beginning of the nineteenth century, it must have seemed as if there was as much whisky coming out of Speyside as there was water in its rivers. This was the height of the smuggling era, and the region's woods and hidden valleys were perfect for hiding illegal stills. There were around 200 stills in Glenlivet alone at this time.

The end of smuggling

It was a Speyside aristocrat, Alexander Gordon, fourth Duke of Gordon, who initiated the change in law which brought the smuggling era to an end and ushered in the modern era of malt. He argued that if distillers were allowed to take out licences and sell their whisky on the open market, then he and the other landowners would help to put a stop to illicit distilling.

The first person to make this shift from moonshiner to legal distiller was George Smith, who had previously been making whisky on his farm in Upper Drumin. His change of status didn't exactly endear him to his neighbours, who threatened to burn down his Glenlivet distillery, and for years Smith always wore two loaded pistols to protect himself from the vengeful smugglers. By then the government, aided by landowners, had put paid to the illegal distillers' resistance. By 1836 the majority of smugglers were either legal whisky-makers, or had returned to farming.

Royal patronage

Glenlivet had already built a considerable reputation for itself prior to the change in law. It was a favourite of King George IV, who asked for it when he visited Edinburgh in 1822. This was slightly embarrassing as, under the laws of the time, it wasn't legal to sell whisky made above the Highland Line in the Lowlands. But George got his way (as kings always do) and Speyside began to build on this patronage.

George Smith was soon joined by other gentlemen distillers and Speyside's first blossoming was underway, with many of the new distillers trying to pass their whisky off as Glenlivet. In time, there were so many Glenlivets on the market that it became known as the longest glen in Scotland. In 1880, Smith's son took

action to protect the apellation. The compromise that was reached gave Smith the sole right to use the title The Glenlivet, but other Speyside distilleries were allowed to affix the designation after their name. Until recently, up to 20 distilleries, including Macallan, Glenfarclas and Balvenie, continued to do this.

This was still a remote region last century. There were no roads: you had to use packhorses to get the whisky to the sea ports. It was only once James Grant (of Glen Grant fame) brought the first railway line into Rothes that Speyside became properly linked with the south of Scotland. This not only allowed distillers to get their malts to the blending centres in Perth, Glasgow and Leith, but enabled coal from the Lowlands to be brought up to fuel the maltings and stills – and therefore increase production.

That said, Speyside still clung to the old ways. Sir Robert Bruce Lockhart's marvellous book *Scotch* contains an evocative recollection of his boyhood in the region at the beginning of the century. In those days it was a quiet area where fishing was free, landlords were benign and Gaelic was still widely spoken. These days,

LEFT

AS FINE A RANGE AS ANY

BELOW

THE ONE THAT KICKED IT

ALL OFF

RIGHT

A MALT WITH CONSIDERABLE
DEPTH

OPPOSITE

ALCHEMY AT WORK AT
MACALLAN

it's full to the brim with whisky tourists and you're more likely to hear Japanese or German than Scotland's native tongue.

By the end of the nineteenth century, blended whisky was the world's favourite spirit. That meant the blenders needed increasingly large amounts of malt. Speyside was perfectly positioned to take advantage – not only could the whisky get to the blending centres quickly, the first master blenders recognized that it produced hugely impressive malt which gave complexity and added sophistication to a blend. Even today there are few top blends that do not have a clutch of Speysides as their core malts.

There's no real surprise, then, that when malt whisky was finally taken out of the hands of a few aficionados and placed onto the world stage, it was a Speyside malt that led the way – Glenfiddich.

The ideal ingredients

Speyside was (and is) blessed with the perfect components for whisky making. It's adjacent to Scotland's main

barley-producing region – the Laich O'Moray – there was peat from the Faemussach Moss, abundant water, years of expertise and transport links.

The water is ideal for making whisky. It has picked up some acidity from the layer of peat that blankets the granite hills, it's iron-free and contains trace elements such as calcium and copper. Virtually every distillery has its own water source and the tiny variations between them are a fundamental difference to each malt.

There's little peat used here these days – although the situation was different at the end of last century. That said, Seagram experimented by giving one of its malts, Glen Keith a heavy peating. The resulting whisky, Glenisla (pun no doubt intended), is a tremendous dram which, when it was shown at a recent nosing, had everyone in attendance clamouring for the firm to start making it on a commercial basis. Although that, sadly, is unlikely to happen, the Glenisla experiment does demonstrate how some malts bloom in the presence of peat. In normal circumstances, Glen Keith is a pretty bland dram; with peat it is multi-dimensional.

There again, trying to get hold of just what it is that makes each whisky unique is as easy as trying to catch a Spey salmon with your fingers. Every distillery manager will have his (or her) own opinion on what the magic ingredient is. Some place the emphasis on the water, others talk up the strain of barley, some insist it's the stills that make the real difference, while others point to the type of wood used.

Stills in all shapes and sizes

Take the stills, for example. You've never seen such a weird collection of shapes in your life. They are enormous (Glenfarclas) and minuscule (Macallan and Glenfiddich), there are some that are heated with steam, with gas or with flame (like Macallan, Glen Grant, Glenfarclas and Glenfiddich), some have got reflux bulbs, others have purifiers (Glen Grant has both). The

lye pipes go up, down and sideways, at Cragganmore
you've even got a flat-topped still, while each still in
Mortlach is a different shape and size. Is there any surprise, then, that we're dealing with a host of different
flavours here?

Wood, and in particular ex-sherry casks, also has its
part to play in the unique character of Speyside's malts.
There aren't many malts that can cope with sherry – it
tends to swamp light, estery flavours – but there are
some rich, complex spirits flowing out of Speyside's

The Speyside divide

So is there such a thing as an identifiable Speyside style? It may be logical to group malts into regions, but it's often the case that they bear little resemblance to each other. Regionality becomes more of a topic in Speyside simply because of the large number of distilleries. Surely there must be similarities between them? The trouble is that there are so many it's difficult to find any sort of overall pattern.

There have been some novel ways of trying to get to grips with this dilemma. Tim Fiddler, in the *Scotch Malt Whisky Society Newsletter*, proposed eight clusters of distilleries that can be compared and contrasted. He divides Speyside into the Banffshire Bank, the Morayshire Bank, then the distilleries in each of the following towns: Rothes, Dufftown, Elgin, and Keith. The remaining two groupings take in the distilleries on the coast and the eastern hills.

If this seems a little excessive, then try the compromise approach taken by Chivas Brothers, which is to split Speyside into north and south. The coast, Elgin, Keith and Rothes are in the north, and Dufftown, the Banffshire and Morayshire Banks and the upland distilleries are in the south. This gives you two roughly equal groups. While each malt has its own personality, you can generalize and say that the Northern Speyside group has greater complexity and sweetness than the drier more malty band from the south.

That said, one of the joys of Speyside is having the opportunity to walk from one distillery to another and compare and contrast the two malts. Tamdhu and Knockando, for example, are opposite each other but make completely different whiskies. Glen Grant and Caperdonich have the same water source, but you'd never confuse one with the other. They may have a family resemblance — the elusive Speyside signature — but they are not copies of each other.

ABOVE
ABERLOUR: AMONG THE GREATS

stills that revel in its rich, ripe fruit flavours. Macallan only ages in Spanish sherry wood, Glenfarclas is almost all sherry these days, while the recent upping of the sherry component in Aberlour has elevated this good dram into a great one.

ABOVE
GLENFIDDICH'S STILLS ARE
AMONG THE SMALLEST IN
SPEYSIDE

Top Speyside malts

Right enough, maybe we shouldn't be that surprised about the number of top-drawer Speysides. If you've got close on 60 distilleries, you'd rightly expect that a number would be at the top of the tree. The rewarding thing is how many of them are truly world-class. Choosing the distilleries that make up the Speyside Classics (see pages 46–49) was a very tough assignment.

They range from the big, boisterous, richly-flavoured examples like the supremely elegant, rich Macallan to the magnificent selection from Glenfarclas, which gain in weight and style the older they get. Light and slightly grassy as a 10-year-old, Glenfarclas picks up flavours of mint, fruit and nuts after a further five years in sherry wood. By the time it's reached 30 years of age it has turned into one of the greatest malts in Scotland – and one of the finest spirits in the world.

Then there's also creamy, elegant Aberlour which, after being hidden from most people's view for years, now seems to be releasing a new version every week. There's a cracking cask strength, a sherry finish and a 100 percent sherry wood to choose from.

Cragganmore, despite being one of UDV's Classic Malts, is little talked about – a baffling state of affairs since it represents Speyside's complexity at its best. Rich, slightly smoky, with a lick of sweetness in the middle of the palate, this is a malt you can roll around the mouth for ages and never quite capture all of the flavours. Equally subtle is the range from Glenfiddich's sister distillery Balvenie, or the fresh and delicate Knockando or the beautifully balanced Tamdhu.

There are also malts that are often dismissed without a second thought. Glenlivet is one that's well worth re-examining, as is Glen Grant, which is often castigated for being light and inconsequential – mainly because it's sold as a five-year-old in Italy. Give it some time in wood, though, and astonishing things happen. Glen Grant is one of those malts that benefits from long ageing – in fact, it continues to develop when other malts are holding up their hands in submission. On first tasting you think it's a pretty, easy-drinking drop, but under that attractive facade is a surprisingly tough, complex core.

Brand or malt?

Then there's Glenfiddich. These days the biggest-selling malt in the world has outgrown the category it established back in the 1970s. It's less of a malt now, more a brand. In fact its owner handles it more as a premium whisky that competes with the likes of Johnnie Walker Black Label or Chivas Regal. That's no bad thing, because the standard bottling doesn't have much in common with the glorious malts that Speyside can produce. It may make sense commercially, but the end result has been that malt aficionados dismiss the distillery out of hand. It's a pity, because when Glenfiddich is aged for a decent time – say 18 years – it reveals its true colours as an elegant balanced dram, though the current vatting leaves much to be desired.

That still leaves four other distilleries that are criminally passed over: Mortlach, Strathisla, Longmorn and Linkwood. Mortlach, like so much of the UDV stable, has remained lurking in the shadows. Rarely seen, talked about in reverent tones by those fortunate enough to have discovered its charms, it's not so much big as massive. A powerfully-built, broody, intense malt, it shares some characteristics with the weighty, rich Longmorn. Both are as solid and enigmatic as the bulk of Ben Rinnes, the gloomy mountain that dominates the Speyside landscape. Strathisla and Linkwood, meanwhile, are as elusive as the Spey itself, charming, peaceful but with surprising hidden depths.

Unrivalled quality

At the end of the day the reason that Speyside has endured while other regions have fallen by the wayside is because its whiskies, on the whole, have a quality that no other region can rival. While other areas rely on one or two dominant characteristics to give them their identity – sweetness in Perthshire, heathery aromas in the north, peat smoke and ozone from the islands – Speyside has a vast range of balanced flavours to its name. It remains malt's glorious, golden heart.

SPEYSIDE

THE CLASSIC MALTS

SPEYSIDE HAS BEEN CALLED THE "GOLDEN TRIANGLE" OF MALT. CERTAINLY NO OTHER REGION HAS SUCH A CONCENTRATION OF HIGH-QUALITY MALTS. COMPLEXITY AND MARVELLOUS BALANCE IS THE ORDER OF THE DAY HERE. THOUGH THEY SHARE A CERTAIN HIGH-TONED SOPHISTICATION, YOU'LL FIND INFINITE VARIATIONS BEING PLAYED ON THIS THEME. THERE ARE SOME WHICH REVEL IN THE LAVISH, RICH NOTES GIVEN BY SHERRY WOOD, OTHERS WHICH REMAIN CLEAN, "AIRY" AND ALWAYS UTTERLY ELEGANT.

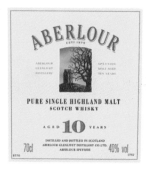

ABERLOUR

10-year-old (40% ABV)

Clean nose with toffee and vanilla notes. Rich and smooth with a clean drive.

ABERLOUR

15-year-old Sherry Finish (40% ABV)

Creamily sweet nose with some butterscotch. Rich and luscious with a raisin/sultana fruit centre and a wheat-cracker finish.

ABERLOUR

18-year-old Sherry Matured (43% ABV)

Elegant nose with plummy fruit. Sweet, rich and chewy with a very long rounded finish.

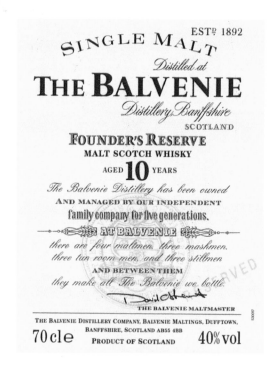

BALVENIE

10-year-old (40% ABV)

A slightly shy aroma with citrus, herbs and a little smoke. Sultry, sweet and elegantly balanced.

BALVENIE

Double Wood 12-year-old (43% ABV)

A similar herbiness, but with richer consistency.
Tangerine, butter and nuts on the palate. Great poise.

GLENFARCLAS

105 (60% ABV)

The alcohol is a little too aggressive, but there's a juici-
ly sweet core lurking behind.

CRAGGANMORE

12-year-old (40% ABV)

A complex, perfumed nose that melds incense, plums
and smoke. The palate starts in a malty fashion that
leads to a sweet centre with sloe berries and smoke.
Rich and superbly balanced. Quintessential Speyside.

GLENFARCLAS

30-year-old (43% ABV)

Sweet, almost raisined/pruney nose with a touch of
rancio – sweet fruits, nuts and cheese. Roast almonds,
walnuts and Seville orange-accented fruitiness on the
palate. Powerful, everlasting finish. A malt to kill for.

CRAGGANMORE

1984 Port Finish (40% ABV)

Sweet, rich red fruit nose. Rich and deep with a
plummy depth. Multi-layered, long and round.

GLENFIDDICH *40% ABV*

Very pale with a very light, sweet nose. Simple.

GLENFIDDICH

18-year-old

Pale colour with a light, brackeny nose and a sweet, clean palate.

GLENLIVET

12-year-old (40% ABV)

A clean nose with plenty of creamy, floral fruitiness and a wisp of smoke. Light and heathery on the palate. Fruity and crisp all the way.

GLENLIVET

18-year-old (43% ABV)

Rich, lightly smoky nose with notes of peel, nutmeg, honeyed fruit and some hickory notes. It eases into the mouth with smooth, elegant sherry nuttiness, cream and a kick of peat. Long and smooth.

GLEN GRANT

10-year-old (40% ABV)

Mealy and floral with some crisp apple aromas. With water, strawberries and cream emerge. Mid-weight and crisp, but with surprising substance.

KNOCKANDO

1980 (43% ABV)

Light, airy nose with gentle blossom aromas. Flowery and with hints of Greek yoghurt and honey behind the crisp palate. A gracious lunchtime dram.

LINKWOOD

12-year-old (43% ABV)

Zingy nose with lemon, sandalwood and some alcohol. A lightly herbaceous start, clean and crisp.

LONGMORN

15-year-old (45% ABV)

Hugely ripe fruit with some treacle, toffee and honeycomb. On the palate there are molasses, dried herbs and nuts. Unctuous, elegant and dense.

MORTLACH

16-year-old (43% ABV)

Firm and rich nose with a rugged earthiness and pent-up potential. Ripe and concentrated exotic aromas of incense, caramelized fruit. A massive, powerful dram that's rich but not sweet.

MACALLAN

18-year-old (43% ABV)

Rich and well-sherried nose reminiscent of Dundee cake, walnuts and ripe, plummy fruit. A rich, complex nose with tablet and chocolate. A crisp, floral mid-palate stops it being cloying. In a different class.

STRATHISLA

12-year-old (43% ABV)

Rich with good depth. Soft, bracken aromas with complex layers of toffee and toasted hazelnuts. Elegant palate with subtle smoke and creamily layered plum and date fruit.

MACALLAN

Gran Reserva (40% ABV)

Complex nose of pink grapefruit, butterscotch, honey, walnut and a hint of clove. Surprisingly delicate given its dark colour. Fresh with deep fruit. Elegant and long.

TAMDHU *40% ABV*

Crisp, crunchy nose with some grass and freesias. A hint of smoke and mint on the palate. Floral with a clean finish.

THE HIGHLANDS

GEOGRAPHICALLY, THE HIGHLANDS OF SCOTLAND LIE ABOVE A FAULT LINE THAT RUNS DIAGONALLY FROM GIRVAN IN THE WEST TO STONEHAVEN ON THE EAST COAST. FOR A TIME, THIS WAS ALSO THE BOUNDARY OF THE INFAMOUS HIGHLAND LINE. IN AN ATTEMPT TO ERADICATE SMALL-SCALE DISTILLING POST-CULLODEN, THE BRITISH GOVERNMENT DECREED THAT ALTHOUGH WHISKY COULD BE MADE IN THE HIGHLANDS (IN A DESIGNATED SIZE OF STILL AND WITH LOCAL INGREDIENTS), FROM 1784 TO 1816 IT COULDN'T BE SOLD BELOW THE LINE. NEEDLESS TO SAY IT WAS. IN FACT, THE SMALLER STILL AND THE EXCLUSIVE USE OF BARLEY MADE HIGHLAND WHISKY A HIGHLY DESIRABLE DRINK, ALTHOUGH IT WASN'T UNTIL THE LICENSING OF DISTILLERIES AND THE COLUMN STILL THAT COMMERCIAL QUANTITIES OF HIGHLAND WHISKY WERE MADE, MOST OF WHICH ENDED UP IN BLENDS.

For simplicity's sake, the Highlands have traditionally been divided into four. South and Perthshire; West; East and Aberdeenshire; North. It makes assessing easier – and also provides the malt lover with some intriguing comparisons.

Scotland is a collection of mini-lands banded together by river, mountain ridge and loch. First-time visitors will not only be surprised at often finding four seasons in one day, but in the 170-mile drive from

Glasgow to Inverness you'll pass from flat, alluvial plains through soft hills and moors and then into wild areas with no sign of habitation. If walking is easy in the southern hills, by the time you reach Dalwhinnie distillery, the 3,000 foot summits of the Monadhliath and Cairngorm mountains are lowering around you – wild

and beautiful tundra country, lit gold and blue by the sun. The malts themselves offer a parallel journey.

South and Perthshire

Most of the distilleries in this grouping owe their existence to the emergence in the nineteenth century of blended Scotch. Although a huge number of new sites were built, many were upgraded farm distilleries. Transport to the blending centres of Perth, Glasgow and Leith was easy, peat was only used lightly (after all, coal was plentiful) and a lighter style soon evolved.

Although few of these nineteenth-century plants have survived, this refreshing, slightly sweet style can be seen in malts such as Glengoyne, which markets itself as Scotland's unpeated malt. This charming old farm distillery is within sight of Glasgow's north-western suburbs, but lies at the foot of the first band of low hills that straggle along the Highlands' southern border. In fact, Glengoyne's warehouses, although just across the road from the distillery, are technically in the Lowlands! Deanston, another of the malts from this area, shares this discreet, soft, unpeated style.

Once north of Perth the hills become higher and though the valleys are still broad and lush, there is a feel-

BELOW
PICTURE PERFECT:
GLENTURRET

ABOVE

SAMPLING AT DEANSTON
DISTILLERY

ing that you are now in the ante-chamber of the Highlands. Traditionally, the emphasis has been on sup-plying malt (fillings) for the blends – no surprise that Aberfeldy and Blair Atholl were built by Dewar's and Bell's respectively. You can find single malt bottlings of both – if a choice is to be made, go for the Aberfeldy.

Glenturret – a beautifully situated distillery – hard-ly needs to sell on the open market. Up to 250,000 vis-itors parade through its doors every year, the bulk of them buying one or more bottles. The amazing success of Glenturret as a tourist attraction has made its owners

a little too keen to sell it at every age and strength known to man – and even as the base for a liqueur.

All these malts share a common Perthshire character of soft, slightly honeyed aromas combined with smidgens of smoke and a dry finish. Pleasant enough, but not a patch on the malt that comes from Edradour, Scotland's smallest distillery. This is distilling as it used to be. Hidden up a glen, out of sight of the road, next to a fast-running burn, using minuscule stills, the end result is the quintessential Perthshire dram, flowery, spicy, lay-ered with honey and cream.

The last of these southern malts is Dalwhinnie which lies, splendidly isolated, close to the top of the Drumochter Pass. The scenery has changed from expansive vistas of mountain and glen to a feeling of being crushed between the massive bulwarks of peaks that loom on either side of the road. It's a crossroads – between two types of landscape and between the soft malts of the south and the more heathery, peat-accented drams that thrive in the north. A middle ground, perhaps, but not a compromise.

West

If the southern malts form an easily-followed trail, then you have to be prepared to travel large distances to take in the distilleries on Scotland's west coast. The small distilleries that thrived among the crofting communities of this, the most spectacularly beautiful part of Scotland, have long disappeared; as have those which sprang up on the shores of Loch Fyne and the Clyde estuary, whose whisky would have been taken by boat or by train to the blenders in Glasgow. Now there are only two left.

Oban looks out to the town's pier, filled with the ferries sailing to and from the Western Isles, and its malt has more of an island character. Some 50 miles north, in the depressing town of Fort William, is Ben Nevis. Owned by the Japanese firm Nikka, most of its production finishes in Japanese brands, which is a shame, for the single malt is a bruiser, filled with rich, concentrated flavours. In many ways it's like the mountain from which it takes its name, massive, apparently impenetrable, but worth exploring.

East

Perhaps it's just the power of suggestion, but Ben Nevis shares this rich character with another mountain malt, although this comes from the eastern extremity of the Grampians – Royal Lochnagar. It's a fine representative of what is now an endangered species, malts from the eastern Highlands. At one time these could have quite easily formed two exclusive groups – those from the valley of Strathmore and the Mearns, the pasture land that runs inland from the coast to the eastern limits of the Grampians – and the rest, from the hills and fertile valleys in central and northern Aberdeenshire.

Now, sadly, there are only four distilleries in operation: Glencadam in Strathmore; Old Fettercairn in the Mearns; Glen Garioch in the Aberdeenshire town of Oldmeldrum; and Royal Lochnagar in the heart of Deeside. The first two are eminently forgettable, but Royal Lochnagar is a stunner – while Glen Garioch is an oft-overlooked charmer. Hailing from the eastern end of the Garioch valley, it's part of the Morrison Bowmore stable, so little surprise that it has floor maltings and also uses a fair whack of peat. Herbal and highly aromatic, it shows that this part of the country can produce characterful drams which are more than capable of coping with the influence of sherry wood. It has only recently reopened, unlike two other distilleries from further west in the Garioch – Glendronach, from Huntly, and Ardmore, whose single

LEFT

SMALL, BUT PERFECTLY FORMED

BELOW

PERFUME AND PEAT IN ABERDEENSHIRE

malt (often at a similar peating level to Glen Garioch) is a rarity.

Most of the other eastern distilleries have either been bulldozed or converted. You may be able to find the occasional bottling of brands like North Port (from Glencadam's home town of Brechin) and Glenurie Royal (from the seaside town of Stonehaven), but stocks are running out. There could be better news for another coastal distillery, Montrose's Lochside, but apart from that, there's little hope that the east will rise once more.

North

What of the north then? Well, there are currently seven distilleries operating in the coastal fringe to the north and east of Inverness – although sadly there's nothing in the capital of the Highlands to show that it, too, used to produce a fair amount of whisky. The Black Isle, which lies between the Moray and Cromarty firths, was home to Scotland's first whisky brand Ferintosh, a drink which was granted duty-free status thanks to the political allegiance of its owner after the first Jacobite uprising. Ferintosh is long gone; the only distillery close to it is Glen Ord, home not only to a massive maltings – which produces malt for most of UDV's Highland and island distilleries – but a medium-bodied, slightly smoky, sherried malt that's gathering a considerable following.

The rest of the northern distilleries are situated on the coast, a location which lends most of them some maritime character, a whiff of ozone on the part of Glenmorangie; salty notes in Pulteney and Balblair; to the kelpy, drying sea-shell aromas of Clynelish. Although the group contains malts that range from the delicate, aromatic Glenmorangie to the ruggedly handsome Clynelish and Balblair, they can be linked with a thread of heathery, sometimes earthy, aromas and flavours that often contain notes of wild herbs. The honey and cream of the southern Highlands has given

way to something harder. They can stand up to high peating levels – the old Brora distillery was used to produce a (very fine) Talisker-style malt when the Skye distillery was gutted by fire – but tend to shy away from heavy use of sherry wood.

That latter quality hasn't prevented Glenmorangie, which remains Scotland's top-selling malt, from undertaking an in-depth analysis of wood management to see which type suits its malt the best. While extended sherry maturation is out of the question, there's nothing to stop the distiller from ageing the malt first in ex-Bourbon casks and then giving it a short period of finishing in other types of cask. The fascinating, complex range of Glenmorangie "Finishes" has prompted other distillers to try similar experiments – and a new mini-category has been born.

This region was the scene of the most brutal clearances on the Scottish mainland. In a final bitterly ironic twist, the landowners who perpetrated this ethnic cleansing, most notably the Duke of Sutherland, then started distilling with the crops grown on their tenants' old land. This wild, often lonely, country sits in mute memory to them.

THE HIGHLANDS

THE CLASSIC MALTS

IT'S UNFAIR TO LUMP EVERY MAINLAND MALT NOT IN SPEYSIDE OR THE LOWLANDS INTO THIS CATCH-ALL CATEGORY. THE SOUTHERN HIGHLAND AND PERTHSHIRE MALTS HAVE A HONEYED CHARM, WHILE THOSE FROM THE EAST COAST SHOW A DISTINCTIVE MALTY, SUCCULENT EXOTICISM. THE WEST COAST IS VIRTUALLY DISTILLERY-FREE, BUT ONE OF ITS TOP MALTS SHOWS A DISTINCT SEASIDE AIR — A QUALITY THAT DOMINATES THE AROMAS OF THE MALTS FROM THE FAR NORTH, A REGION WHOSE RANGE OF COMPLEX, MULTI-FACETED MALTS IS OFTEN OVERLOOKED.

CLYNELISH (NORTH)

14-year-old (43% ABV)

Filled with seashore aromas: rugged, with a hint of kelp at the high water mark. Smoky start with an iodine tingle, before a sweet mid-palate with some sultana fruitiness and a dry finish.

DALMORE (NORTH)

12-year-old (40% ABV)

A big, bracing, slightly salty aroma with blackcurrant sweetness behind. Rich, oily start then the complex flavours atomize across the palate, with black fruit dominating.

DALWHINNIE (CENTRAL)

15-year-old (43% ABV)

Light, slightly malty nose. The palate balances honey and pepper well.

EDRADOUR (CENTRAL)

(40% ABV)

Rich, evocative nose with hints of silage, waxed rain-coats, sawdust, herbs and wild flowers. Ripe, luscious palate with silky-smooth fruit and a lick of heather honey on the finish.

GLEN GARIOCH (EAST)

15-year-old (43% ABV)

Highly intense, perfumed lemon/ginger nose with some floor polish and peat. The smoke is more pronounced on the palate.

GLEN ORD (NORTH)

12-year-old (40% ABV)

Excitingly fresh, malty nose where the sweetness is checked by a lime leaf tingle. An ever-changing palate with a hint of smoke, citrus fruits, sweet fruit and a dry, roasted nutty finish.

GLENGOYNE (SOUTH)

17-year-old (43% ABV)

Some almond, lightly-sherried fruitiness on the nose along with calf leather, hazelnut and clover. Well rounded palate. A perfect lunchtime dram.

OBAN (WEST)

14-year-old (43% ABV)

Clean, tingly nose filled with dried herbs and sea breezes. A softly spicy, slightly smoky palate balanced with refreshing saltiness.

PULTENEY (NORTH)

12-year-old (40% ABV)

An ozone-fresh nose that also has attractive cakey notes. Starts sweet, with some sugared almond, then menthol and cigar-box flavours. A complex beauty.

ROYAL LOCHNAGAR (EAST)

23-year-old (59.7% ABV)

Sherry sweet on the nose. A hugely, muscular, sturdy number with good spicy notes. Rounded, rich and mellow, it copes with its high strength easily.

GLENMORANGIE (NORTH)

10-year-old (40% ABV)

Delicate, pear-drop nose. Fresh and grassy. Smooth, creamy, gentle and approachable.

BLENDED SCOTCH

BLENDED WHISKIES HAVE, IN RECENT TIMES, BEEN SEEN AS THE ALSO-RANS OF THE SCOTCH. THE FACT REMAINS THAT 95 PER CENT OF THE SCOTCH WHISKY SOLD IN THE WORLD IS BLENDED. ALL THESE DEDICATED DRINKERS CAN'T BE WRONG, BUT WE ARE NOW IN A WORLD THAT APPARENTLY BELIEVES THAT THE INDIVIDUAL IS OF GREATER WORTH THAN THE COLLECTIVE. BLENDS ARE SUFFERING. THEIR SHARE OF THE INTERNATIONAL WHISKY MARKET IS SLIPPING, NO ONE WRITES ABOUT THEM WITH ANY SERIOUSNESS, AND THEY ARE AN ANACHRONISM IN TODAY'S SINGLE MALT WORLD.

RIGHT
TOMMY DEWAR: GREATEST OF THE WHISKY BARONS

Don't write off blends

Well, before they finally disappear from consideration, let's have a look at how blends put Scotch whisky on the world stage, and how their assembly is every bit as creative an exercise as the making of a malt.

Scotland in the early part of the nineteenth century was a country in flux. The Highlands were being brutally cleared to make way for large tracts of land for sheep, initiating a mass exodus from country to town – and to America, Canada and Australia. The population became concentrated around the new centres of heavy industry in the coal-rich Lowland belt.

The arrival of heavy industry put central Scotland at the forefront of the industrial revolution, with Glasgow as its engine room, the furnace that fuelled the growing power of the British Empire. But Glasgow's fires ran not only on coal, but on people's sweat, and they needed something to slake their thirst.

The Lowlands had long made whisky, but by and large it was pretty poor stuff. While Highlanders concentrated on making small batches of malt to sell or drink in their own communities, Lowland distillers had always to serve a much larger market – and that meant building bigger stills, running them as fast as they could and using whatever grain which came to hand.

A trade in alcohol had long been established with England in which Lowland whisky would be sent south to be rectified (redistilled) and turned into gin. We are already therefore talking about industrial-sized volumes. The influx in population meant that production had to be increased once more, but the whisky that came off the stills was pretty foul stuff. Clearly something needed to be done to satisfy the thirst of the growing urban masses.

LEFT
RABBIE BURNS LIKED
A DROP ...

The Coffey still

In 1826, Robert Stein, cousin of John Haig, a member of the family which dominated Lowland distilling at that time, appeared with a revolutionary new still which Haig promptly installed in his Cameronbridge distillery.

The theory behind this was that as long as you put wash in one end you could get high-strength spirit out the other, with no need for a second distillation. It worked, but needed some tweaking, which was done in 1830 by the former Excise General of Ireland, Aeneas Coffey.

Coffey's patent still was a godsend to the Lowland distillers, all of whom soon had one working in their distillery. The whisky it produced was high-strength and light in flavour, which suited the English gin distillers, but it wasn't quite to the taste of the Lowland Scots.

Then fate took charge. The Wash Act of 1823 had created a rash of newly legal malt distilleries, which had expanded their production. Though malt was highly regarded by the cognoscenti, it was too heavily flavoured for most people in the Lowlands and in England.

The birth of blends

Then some bright spark had the idea that if they combined a small amount of the richly-flavoured malts with the lighter grain, they'd come up with a commercially acceptable middle ground. Officially, blends were born in 1853 when Usher's Old Vatted Glenlivet appeared in Edinburgh. The floodgates were opened.

The majority of the early blenders were high-class grocers and wine merchants based in the growing centres of population. John Dewar, Arthur Bell and Matthew Gloag were all in Perth; James Chivas in Aberdeen; William Teacher in Glasgow; George Ballantine in Edinburgh; John Walker in Kilmarnock. All were well versed in blending teas. The principle of blending whiskies was much the same.

That, of course, didn't guarantee success. For that it needed business skills previously unseen in the British drinks trade. The sons and grandsons of the whisky blenders weren't just concerned about selling to their existing clientele, they wanted to sell whisky to the world. Their story is worthy of a book in its own right.

What they did, briefly, was to cajole, bully and charm their way into the bars, hotels and restaurants of the world through a succession of stunts and slightly dodgy business practices. By the end of the century they were shipping millions of bottles worldwide.

This entrepreneurial zeal was undoubtedly helped by the effects of the vine louse phylloxera which halted brandy production in its tracks, by the reluctance of the Irish to use the patent still and by the distribution network of the British Empire which took the blends to the Highland diaspora.

The blenders started building distilleries in Speyside, the Perthshire glens and on Islay to satisfy the growing demand. From then on, the function of the majority of malt distilleries was to provide fillings for blends. The situation remains virtually unchanged today.

The art of blending has remained much the same as well, even though the constituent parts of the blend will also have changed. Many of the malt distilleries that supplied fillings for the early blends no longer exist. The blenders therefore have to find alternatives. But how, if, as legend would have it, every malt has its own taste, is this possible? Welcome to the world of the blender.

What is a blend?

In its simplest sense, a blend is a mélange of grain and malt whiskies assembled to produce a consistent, identifiable brand. But things are never as simple as they seem in the world of whisky. Take malt, for example. Each day's distillate is slightly different, each cask is a distinct individual. The palette changes continually.

Grain whiskies, contrary to popular belief, behave in the same way. Each distillery has its own signature – from clean and light, to round, buttery and fat. Some are more suitable to long ageing than others – and they, too, will vary in aroma from day to day.

When you add in the fact that distilleries close down rather too regularly for comfort – meaning there's a finite supply of some whiskies – you can begin to understand that the blender, rather than working with unchanging blocks of flavour, is dealing with ever-shifting aromas and possibilities.

At the heart of each blend are a number of core malts, usually from Speyside. These give the signature flavour(s) to the final product – at the heart of Johnnie Walker, for example, are Cardhu and Talisker, while Chivas relies on Strathisla. Although vitally important, these core malts should never dominate the blend; in other words, although you may be able to detect a smoky element in Johnnie Walker, you shouldn't be able to say that it smells and tastes of Talisker.

This unchanging core is then buttressed by a layer of supporting malts, each of which adds some other shading of aroma or flavour; and then finally by a larger amount of packers. The function of these is to give the blend body and texture, rather than flavour. This is the interchangeable part of blending. The packers can come from a large number of distilleries, and the blender can switch from one packing distillery to another without affecting the overall character or flavour of his blend.

ABOVE

A LIGHTER STYLE: THANKS TO THE BARRELS ...

RIGHT

RICHER: THANKS TO SPANISH WOOD

At the end of the day there are some malts that are more highly prized as blending whiskies than others. Malts like Aultmore, Mortlach and Cragganmore are in the top league – one reason why the first two are rarely seen as single malt bottlings. All share a complexity of flavour and a richness of texture that melds well with other malts and grains. There are other fine single malts – Glen Grant, for example – which are awkward customers in a blend. What the master blender is looking for is a nucleus of malts that complement each other and bring out different aspects of each other's personality.

Some blenders use malts from each of the regions, others rely on one – Bell's is a country-wide blend, Chivas and J&B are exclusively from Speyside, Black Bottle is predominantly from Islay. Much, to be honest, depends on how many distilleries the blender controls. It's often thought, therefore, that the higher the number of malts in a blend the better it is. That isn't necessarily true. It's the quality of the whiskies and the nature of their interplay that matters.

Wood flavouring

The final element in building flavour and character is the type of wood that the malt has been aged in. A light blend like Cutty Sark will use predominantly refill, while the richer Famous Grouse will have malts that have been aged in ex-sherry butts made from Spanish oak.

Age is important as well. Different malts (and grains) will be used at different times. A single malt bottling will ideally show the malt at its peak; a blender, however, may want to show a slightly different aspect. A mass of possibilities exists for the blender.

Then there's grain. It, too, undergoes the same process as malt – a central grain at the heart, supported by others, each of which brings its own personality to the party. Finally all the components are brought together, but before blending they are nosed one final time and adjustments are made – some malts may be upped, others dropped, substitutes brought on. The final bottle may taste the same as ever, but the components may be slightly different.

Today's whisky market may be finally allowing malts to shine, but don't ignore the great blends. Take some time and look at the hugely impressive Chivas range, reacquaint yourself with Ballantine's, J&B, Johnnie Walker and Grouse. The best can rival many malts for complexity and subtlety.

ABOVE

THE BENCHMARK STILL

LEFT

UP AMONG THE BEST

BLENDED SCOTCH

THE CLASSICS

IN THESE TIMES WHEN MALT WHISKY IS RECEIVING ALL THE ATTENTION IT'S EASY TO FORGET THE SHEER QUALITY OF THE GREATEST BLENDED SCOTCHES. THESE WHISKIES, AFTER ALL, WERE THE BRANDS WHICH HELPED SCOTCH BECOME THE WORLD'S FAVOURITE SPIRIT AND THE BEST OF THEM DESERVE THE SAME ATTENTION AND RESPECT AS SINGLE MALTS. "APPROACHABLE", "EASY-DRINKING" AND "VERSATILE" ARE TOO OFTEN THOUGHT OF AS NEGATIVE TERMS. GIVE BLENDS ANOTHER CHANCE – YOU'LL BE SURPRISED.

BALLANTINE'S GOLD SEAL
12-year-old (43% ABV)
A gentle, sweet, chewy nose with a touch of heather, and a lovely, pure, grainy centre. A lush, plush dram.

BALLANTINE'S
18-year-old (43% ABV)
Despite its age, this has retained a clean, grassy nose with well-balanced wood. On the palate it's round with chewy, creamy weight and a hint of pulpy fruit. A gentle, smooth finish. Elegant.

BELL'S
8-year-old (40% ABV)
A clean nose with light toasty wood in evidence. A light grassy aroma then yields to good maltiness with a hint of peat. On the palate it's fairly sweet and round with good balance.

BLACK BOTTLE
10-year-old (40% ABV)
Richly coloured. Pungent peatiness mixed with fresh turf and some coconut matting. The palate is pure and sweet with a refreshing peppery needle and is very well balanced. The finish is long, rich and smoky. One for Islay fanatics.

CHIVAS REGAL

12-year-old (40% ABV)

Light and estery nose with sweet hay and a seductive smokiness in the distance. Clean US wood on the palate and a little touch of sugared almonds mingling with soft, bouncy grass. A gentle and graceful blend.

CHIVAS REGAL

18-year-old (40% ABV)

Big, rich, complex nose with notes of wild herbs, mint, pruney fruit and a light, medicinal character. Robust palate with weighty, almost chocolatey flavours. On the finish there's a touch of heather and lime.

THE FAMOUS GROUSE *40% ABV*

A subtle, smooth and gently smoky nose with hints of sweet sherry wood and mild peatiness. The same character shows on the palate alongside ripe red fruits. A crisp refreshing finish. Well balanced. Still the finest standard blend on the market.

J&B *40% ABV*

Very pale in colour. The crisp nose is of cut straw mixed with some light nuttiness. A creamy note emerges when water is added. A soft, light dram with excellent balance between sweet fruit and a crisp, dry finish.

JOHNNIE WALKER RED LABEL *40% ABV*

Hugely popular – and you can see why. There's a distinct island influence on the nose, along with clean, medium-bodied fruitiness. Clean and easy-drinking.

JOHNNIE WALKER BLACK LABEL *40% ABV*

The nose is instantly seductive, showing the signature touches of peat smoke and rich, sherried fruit character. Big with a malty richness, giving extra texture to the palate. A beautifully complex finish rounds off the benchmark blend. Better than many malts.

IRISH
WHISKEY

FOR A COUNTRY THAT CAN CLAIM TO BE THE BIRTHPLACE OF WHISKY, DISTILLERIES ARE FEW AND FAR BETWEEN IN IRELAND. IN FACT, THERE ARE ONLY THREE IN THE WHOLE OF THE COUNTRY. YET AT THE TURN OF THE CENTURY, IRELAND WAS SEEMINGLY SET FAIR TO RIVAL SCOTCH AS A MAJOR INTERNATIONAL PLAYER. HOW COULD A COUNTRY THAT PRODUCES SUCH DELICIOUS, LUDICROUSLY DRINKABLE WHISKIES BECOME A BIT-PART PLAYER IN WHISKY'S EPIC PICTURE?

Whiskey (sic) is inextricably bound up in Ireland's history. It was likely to have been first produced (for medicinal purposes only) by the scholar-monks of the early Celtic Church, the very men who spread the gospel and learning from Ireland into Scotland and Western Europe at the end of the Dark Ages.

By the twelfth century, uisce beatha was no longer just a medicine but a powerful stimulant – used by war-

lords to give their troops a jolt of courage before going into battle.

Ireland has always had a rural economy and the people, eking out their living on its green fields, supplemented their income with making and selling whiskey. It had a practical as well as a social function – helping to pay the rent, as well as making the fiddler's elbow go that little bit faster. That said, there has always been a split between country (also known as poitin) and town whiskey. Up until the beginning of the nineteenth century much of the town whiskey – the commercial brands – was flavoured with herbs and roots. The spirit

RIGHT

THE POT STILLS AT MIDLETON

BELOW

BUSHMILLS: PRIDE OF THE
NORTH

enjoyed as fine a reputation as Cognac, and in the eighteenth century Ireland boasted 2,000 distilleries.

This flavoured whiskey was sipped in salons in
Dublin, London and Paris, while the poor people of
Ireland supped on the poitin that ran out of their small,
peat-fired stills – until they were clobbered by the same
legislation that fuelled the smuggling wars in the
Scottish Highlands.

Actually, the farmer-distillers didn't disappear, they
just went underground. Even today if you ask in the
right places for poitin you'll find it without too much
trouble. Right enough, whether you'll get a good drop
or not is a matter of luck and how good your connections are. There's some wild and frightening stuff out
there, but there are also decent (if highly potent) examples. After all, Ireland is all about the craic, and the often

farcical trip needed to root out poitin is one that is well worth taking.

Back in the last century though, the major effect of the legislation was to consolidate whiskey-making in the hands of a few town distillers like John Jameson and John Power, whose pot-still whiskies, now no longer flavoured, had a reputation that outstripped that of Scotch. When the Irishman Aeneas Coffey came up with his patent still they saw the whiskey it produced as a crude imitation and refused to use the new technique.

The distillers in the Lowlands of Scotland (among them Jameson's brother-in-law, John Haig) had no such reservations and soon the international markets were being flooded by their new blends. Ireland was caught napping. Then America, the single biggest market for Irish whiskey, declared Prohibition. At the same time, Ireland won its struggle for independence but was immediately hit by a trade embargo that banned sales of Irish goods in Britain and the Empire. In the space of 20 years Irish whiskey was ruined.

Bizarrely, the Irish Government decided to kick the industry when it was already down and stopped distilling during the Second World War. By 1952, Irish whiskey exports amounted to £500,000. Scotch was worth £32.5 million.

By the mid-1960s, it made sense for three of the last four distilleries in Ireland to join forces and make all their whiskey in one plant in Midelton, Co. Cork. When Bushmills, the sole distiller left in Northern Ireland, joined the new Irish Distillers Group in 1973, all Irish whiskey was made by one firm.

In recent times a newcomer, Cooley, has joined the fray and proved to be the spark the industry needed. Now the two firms are vying with each other to show the wide range of styles that Ireland can produce. Irish Distillers insists on triple distillation and unpeated malt, Cooley uses peat in some brands and has no reservations about double distillation. The added choice has led people back to the gentle Irish style.

The white-walled buildings at Bushmills wouldn't

ABOVE
THE ROCKY LINK BETWEEN
IRELAND AND SCOTLAND

look out of place in Scotland. Indeed, from the nearby Giant's Causeway you can make out the blue fuzz of Islay and Kintyre. The whiskey itself is a spirituous link that replicates the basalt strip that surfaces at the Causeway and Fingal's Cave in Staffa. Midleton is the polar opposite of the classic lines of Bushmills. That said, the public are not allowed to see this miraculous ultra-modern distillery; that would ruin the illusion. Instead they are shown round the much more picturesque old Cork Distillers plant, now rather confusingly called the Jameson Heritage Centre. Midleton isn't romantic, but, like Cooley's highly functional plant in Dundalk, it produces some great whiskies.

Bushmills remains rooted in the paradoxical landscape of the Antrim coast, rugged yet soft, gentle yet bracing. What of the rest of Ireland's brands? Has technology replaced the notion that whiskies reflect their place of birth? Maybe. But each brand still uses its own recipe of malted and unmalted Irish barley, its own combination of pot and column still, and its own choice of casks. At the end of the day, Power's is still the juiciest, peachiest, dreamiest dram you could ask for, Crested 10 and Redbreast are as rich and weighty as ever, and Jameson 1780 still hits the mark. The magic has been retained.

ABOVE
A CROCK OF GOLD AT THE END
OF THE IRISH RAINBOW

IRISH WHISKEY

THE CLASSICS

IRISH WHISKEY IS UNDERGOING A RENAISSANCE. THE COUNTRY MAY ONLY BE ABLE TO BOAST THREE DISTILLERIES, BUT WHAT IRELAND LACKS IN STILLS IT MAKES UP FOR IN HIGH-QUALITY BRANDS. THE BEST ARE SOME OF THE MOST DRINKABLE DRAMS YOU'LL EVER COME ACROSS. THEY COMBINE A FRESH DRIVE WITH RIPE, SWEET TROPICAL FRUITINESS. THEY CAN COPE WITH SHERRY WOOD, WHILE (MOST OF) THE SINGLE MALTS HAVE A CREAMY, CLOVER-LIKE CHARM. BE PREPARED TO BE SEDUCED.

BLACK BUSH (IRISH DISTILLERS) *40% ABV*

A full, sweet nose brimming with nutty butter toffee and a hint of sherry wood. A complex, elegant whisky.

BUSHMILLS (IRISH DISTILLERS)

10-year-old Malt (40% ABV)

A gorgeous mix of vanilla ice cream, clover and liquorice on the nose. A malty bite on the palate before a soft, dry finish.

BUSHMILLS (IRISH DISTILLERS) *40% ABV*

Very soft and fruity nose that mixes white pepper and orange. A delicate, slightly toasty palate.

CONNEMARA MALT (COOLEY) *40% ABV*

A similar level of peatiness to Lagavulin, but altogether more aggressive. The rather damp peatiness dominates an assertive citric palate.

CRESTED TEN (IRISH DISTILLERS) *40% ABV*

A weighty mix of freshly turned earth, malt and fruit on the nose. Broad and almost tarry on the palate it coats the whole of the mouth.

KILBEGGAN (COOLEY) *40% ABV*

A mix of camphor and grass on the nose. A middle-weight whiskey that starts sweetly then finishes with a chilli pepper bite.

JAMESON (IRISH DISTILLERS) *40% ABV*

Soft, full and slightly malty nose. Fresh with a green unmalted barley edge combined with light sweetness. Medium intensity, but still a creamy, smooth experience.

LOCKE'S (COOLEY) *40% ABV*

Soft, clean and quite malty with some nutty sherry wood on show. This sweet component carries through the palate to the peppery finish.

JAMESON (IRISH DISTILLERS)
1780 (40% ABV)

A heady mix of ripe fruit and dog-rose aromas leap out of the glass. Creamy, long and soft with a creamy, crisply nutty palate and a magnificently long finish.

POWER'S (IRISH DISTILLERS) *40% ABV*

A noseful of super-ripe, pulp, apricot, peachy fruit. Softly weighted on the palate, but with a refreshingly edgy backbone. A true classic.

AMERICAN WHISKEY

LEFT

DISTILLERIES HAVE COME AND GONE, BUT BOURBON ENDURES

KENTUCKY, THE STATE OF BLUE MOONS, BLUE-GRASS AND SWEET RAIN, IS AMERICA IN MINIA-TURE. ELEGANT WHITE PALLADIAN MANSIONS, THOROUGHBRED HORSES AND GARDEN PAR-TIES, HEAVY INDUSTRY, DRIVE-IN TATTOO PAR-LOURS AND TOUGH WORKING-CLASS BARS. COUNTRY AND CITY SITTING AWKWARDLY BESIDE EACH OTHER. IT'S THE SOUTH, BUT NOT THE DEEP SOUTH; IT DOESN'T QUALIFY AS PART OF THE MID-WEST, BUT DON'T THINK THAT MEANS IT'S PART OF THE UP-TIGHT EAST. KENTUCKY IS, QUITE SIMPLY, KENTUCKY, EXIST-ING IN ITS OWN TIME-FRAME AND PROUD OF ITS OWN TRADITIONS, CENTRAL TO WHICH IS THE MAKING OF BOURBON.

All these elements come together one day in May when the factory workers mix with the landed Southern gentry on Churchill Downs to watch the Kentucky Derby. Although class divisions are rigidly maintained, they are joined together by a love of horseflesh – and of the Mint Julep, the potent summer cocktail that fuels the meet. On this day bourbon regains its role as Kentucky's single greatest invention and America's spirit.

Kentucky's gentle hills and wooded valleys, cut by rapid streams and rills, are perfect whiskey-making country. There's fertile land to plant crops, limestone-rich water, stands of oak for making barrels. Little surprise that when the land was first settled by the white man in the 1770s, whiskey was soon being produced. There's another good reason for this occurrence. South Carolina, Kentucky and Tennessee were initially settled by Scots and Irish emigrants, people who already had experience of turning excess grain into a high-strength spirit in their homelands, many of whom had already been making rye whiskey in Pennsylvania.

The creation of a commercial industry was given a significant boost by a further wave of distillers, fleeing west into Kentucky and Tennessee (neither of which at that point in time had achieved statehood) to avoid the whiskey tax imposed by Washington to raise money after the War of Independence.

OPPOSITE

THE DISTILLATION OF KEN-TUCKIAN SPIRIT: THE DERBY

KENTUCKY

THESE SETTLERS WERE GIVEN FREE LAND IN KENTUCKY ON THE PROVISO THAT THEY WOULD CLEAR THE LAND AND PLANT INDIAN CORN (MAIZE). THIS THEY DID WITH CONSIDERABLE GUSTO – SO EFFICIENTLY DID THEY CLEAR THE OAK WOODS THAT WITHIN A FEW YEARS, THE KENTUCKY WHISKEY INDUSTRY HAD TO IMPORT WOOD TO MAKE ITS BARRELS.

Looking back on those times it seems as if the state was awash in whiskey. As well as there being stills attached to farms producing small amounts to fuel family and community celebrations, and providing welcome extra cash, larger commercial operations were soon in place – Evan Williams arrived in 1783, Jacob Beam in 1795. As for the Revd. Elijah Craig, who is often credited with inventing bourbon, well he existed and he distilled, but he wasn't the central figure that legend has made him. There was no single founding father; instead there was a mass explosion of whiskey-making.

These sizeable distilleries were situated close to the Ohio river. It was clear that, no matter how strong a thirst the people had – and they were considerably heavier drinkers than we are today – more whiskey was being produced than they could drink. The trouble was that it was well-nigh impossible to get large shipments across the Appalachians to the prosperous eastern markets. The only road was the river. So it was that barrels of whiskey stamped with the legend "Bourbon County"

set out on barges, destination New Orleans. From there it was shipped up the east coast, slowly overthrowing rum as America's favourite spirit.

Distilling was carried westward across the frontier with the railroads, and although Kentucky remained the heart of commercial production and whiskey's spiritual home, it was made wherever grain was grown – you'd find distilleries in Virginia, Pennsylvania, Maryland, Indiana, the Carolinas, Tennessee, Missouri and further west. Some of these produced fine whiskey, others were happy to churn out evil stuff to quench the thirst of the workers.

By the mid-nineteenth century, quality control had improved significantly – mostly thanks to the efforts of Jim Crow, whose researches at the Oscar Pepper distillery into sour mashing, charring barrels and distilling improved the whiskey from the better Kentucky distilleries. The combination of this leap in quality and a swingeing tax imposed by Lincoln after the Civil War resulted in a split between the small farmer distilleries, who soon became marginalized as makers of moonshine, and the increasingly large concerns who had now to supply whiskey for the whole of America.

All seemed set fair. America's population was growing rapidly, the distilleries were growing in size and influence and people's thirsts were prodigious. Continuous stills were beginning to be introduced, thereby increasing quantity, while exports were slowly starting up – and the quality of the whiskey was being appreciated in Europe. At the time it must have seemed inconceivable that America wouldn't become the world's biggest whiskey

RIGHT

WHISKEY-LOVERS GIVE

THANKS FOR TURKEY

producer – how could the tiny countries of Ireland and Scotland possibly compete with this massive land? Perhaps the world of whisky would have been considerably different if the inherent conservatism of American society hadn't reared its head just at the time when its whiskey industry was on the brink of becoming a major international player.

The flipside to the growing whiskey industry was the Temperance Movement. The industry ignored it, mocked it and encouraged heavy consumption, thus driving more people (particularly women) into the arms of the teetotallers. American mores have always swung from puritanical fervour to hedonistic excess with dizzying speed. By the turn of this century that pendu-

lum was definitely swinging in favour of Prohibition. Although the Volstead Act, which banned the sale and consumption of alcohol, didn't come into force until 1919 – soon after women's suffrage – many states, including Kentucky and Tennessee, were already dry.

The absurdity of Prohibition effectively killed the industry and let Scotch dominate the world. Even after Repeal in 1933, it had only a few short years to get up and running again before the government shut it down once more, this time to turn distilleries into producers of alcohol for the war effort. By the time it could start making whiskey again in the 1950s, Scotch had the world of whisky in its grip and America was falling in love with vodka. Few distillers could survive in that climate.

These days Kentucky can boast ten distilleries. Before Prohibition there were 2,000. A part of American history has been lost, along with indigenous styles like rye and corn whiskey. Although Wild Turkey and Jim Beam still make small quantities of (tremendous) rye, no-one is making corn whiskey any more. The best bet for finding some of this sweet but powerful style is to search out a moonshiner in the wilds of the Carolinas.

The archives are littered with fading photos of long-forgotten distilleries, fragments of the history of once-famous brands like Green River, "The Whiskey Without A Headache". You can chart its strange rise and fall, from small label to a giant brand – and one which had more ridiculous marketing stunts attached to it than most – soon followed by the closure of its Owensboro distillery and, after a brief revival in the 1950s, the regular passing of the brand rights from one distiller to another. These days the only Green

River most people will have heard of is a track by Creedence Clearwater Revival.

Thankfully, after a lean period that would have driven most distillers insane, the world appears to be willing to look at bourbon seriously. The top end is being exploited like never before with small-batch, single barrels, barrel select and barrel strength becoming the new flagships. Quality rather than bulk is now the by-word.

Even though the industry has shrunk, you can't help but feel when you slip through the wooded land of Kentucky that this is good whiskey-making country. There's something in the air, in the pace of life that lends itself to the art. The distilleries themselves range from the brutally functional Barton and Beam; to the strangely beautiful monolithic red brick piles of Bernheim, Early Times and Ancient Age; the incongruous Spanish-style Four Roses; the bizarre, black-painted Wild Turkey teetering on the edge of the Kentucky River gorge; to the pretty-as-a-postcard Labrot & Graham and Maker's Mark. Sadly, Heaven Hill's plant is now a jumble of mangled metal and scorched earth, all that's left after a warehouse fire got out of control, sending flaming barrels and a river of blazing whiskey downhill, detonating the distillery.

They may make strange-looking bedfellows, but these days they are producing some magnificent whiskeys that deserve to be taken seriously and, most of all, enjoyed. The best Kentucky whiskeys have complexity by the bucketload but they also have a muscular sweet punch that allows them not only to be sipped on their own, but to form the heart of some of the world's great cocktails – the Manhattan, the Bronx, the Old-Fashioned and the Mint Julep. The world would be a poorer place without them.

KENTUCKY BOURBON

THE CLASSICS

AFTER YEARS OF LANGUISHING IN THE SHADOW OF SCOTCH, BOURBON HAS REINVENTED ITSELF AND SHOWN ITSELF TO BE ONE OF THE WORLD'S GREAT WHISKIES. EVERY BRAND HAS ITS OWN RECIPE OF GRAINS AND OFTEN USES A SECRET YEAST STRAIN, BUT THE END RESULT IS A STYLE OF WHISKEY WHICH COMBINES RICH, SMOKY, VANILLA NOTES WITH A SWEET, FRUITY PUNCH. YOU CAN SIP THEM OR USE THEM AS THE BASE FOR SOME OF THE WORLD'S GREAT COCKTAILS … BUT TRY THEM.

ANCIENT ANCIENT AGE
10-year-old (43% ABV)

Rich, elegant and plump nose with some accents of cinnamon and treacle along with cereal notes. A soft, smooth and gentle start with a crisply long finish.

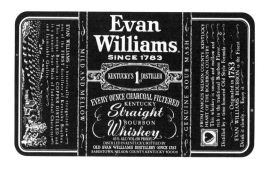

EVAN WILLIAMS (HEAVEN HILL) *45% ABV*

Graceful, if slightly wood-accented nose – spice, cinnamon, caramel and a little smoke. A good rounded bourbon with a mid-palate bite.

ELIJAH CRAIG 12-YEAR-OLD *47% ABV* (HEAVEN HILL)

Rich with spicy wood hinting to nutmeg. A chewy, toffee palate. Great length.

FOUR ROSES YELLOW LABEL *40% ABV*

Gentle, wooded, lightly citric. Clean, delicate and fragrant. A fine mixing brand.

JIM BEAM RYE *40% ABV*

Lemon zest and cumin seed on the nose. Spicy and zingy. A great old style of whiskey that sure wakes those tastebuds up!

VERY OLD BARTON (BARTON) *40% ABV*

The nose is rich, with beech nuts and juicy fruit. Rich palate with a crisp, zippy lemon finish from the rye.

KNOB CREEK 9-YEAR-OLD *50% ABV*

Rich, sweet, fruity nose – with hints of plum, nuts and hickory smoke. Elegant, rich and very ripe palate.

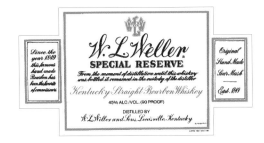

W.L. WELLER (UDV) *45% ABV*

Rich nose with butter, molasses and some rum and raisin. Rich and sweet start with spicy, chocolatey fruit. Soft as velvet.

OLD CHARTER (UDV) *43% ABV*

A complex mix of floral notes and spices – with menthol and lemon balm dominating. Very clean and zesty palate with excellent rounded vanilla fruit.

WILD TURKEY 101° 8-YEAR-OLD *43.4% ABV*

Fine amber colour. Clean, minty nose with great rich vanilla weight. Elegant and rich, brimming with ripe fruits.

TENNESSEE

J UST WHEN YOU THINK THAT YOU'VE GOT TO GRIPS WITH AMERICAN WHISKEY, YOU ARRIVE IN TENNESSEE. THEY DO THINGS THEIR OWN WAY DOWN THERE. IN THE EARLY DAYS, TENNESSEE WHISKEY EVOLVED IN MUCH THE SAME WAY AS THE WHISKEY OF KENTUCKY. THE PRODUCTION PROCESS IS THE SAME AS THAT FOR BOURBON – A MIX OF GRAINS (PREDOMINANTLY CORN, WITH RYE AND MALTED BARLEY MAKING UP THE "SMALL GRAINS" ELEMENT) DISTILLED FIRST IN A COLUMN STILL AND THEN REDISTILLED IN A DOUBLER.

It's widely thought that it's sour mashing that sets Tennessee apart – mainly thanks to the fact it's emblazoned across the Jack Daniel's label. Wrong. All bourbon and Tennessee whiskeys are sour mashed. The difference lies in the use of filtering the new spirit through maple charcoal. This "Lincoln County Process" has been used since the 1820s, giving the state's whiskey an undoubted smoothness and roundness.

So what does the mellowing do? For starters, it extracts the aggressive fusel oils and some high-toned esters. Because vodka producers use the same technique, it is sometimes thought that this means that young Tennessee whiskey is no more than a type of American vodka. Well, it isn't. It's distilled in a different fashion, is lower in strength and uses different grains. While some character will be lost during mellowing, the new spirit has been buffed up and has picked up some sweetness from the maple charcoal. It's different, plain and simple.

These days there are only two distilleries in Tennessee, both drawing water from the same massive underground limestone bench. They couldn't be more different, however. One is the best-known American whiskey in the world, the other is rarely seen.

Jack Daniel's

The history of whiskey-making is littered with remarkable characters, but few can touch Jack Daniel. He was a whiskey-making prodigy who learned distilling at the knees of a grocery store owner and lay preacher called Dan Call who made his hooch at the evocatively-named Louse Creek. Call eventually took the pledge and in 1859 sold the Louse Creek operation to Jack. So

there he was, aged 13, the owner of a distillery.

It didn't seem to faze him. After the Civil War, Jack had already managed to build up a fair amount of money and a good reputation. He therefore needed to expand production. To do that he needed more water. After a short search he found what he was looking for: a deep underground spring in Cave Spring Hollow on the outskirts of the small town of Lynchburg. The scene was set.

If there's one thing that has characterized Jack Daniel's, the brand, it's canny marketing. Today it has managed to meld homespun, folksy advertising with a rebellious image. It is the rock and roll whiskey. Back last century Jack himself was always keen to exploit any marketing possibility.

The distillery was switched to be closer to the rail-road, and when licences were being introduced it was Jack who rushed his application off to Washington in order to become America's first legal distillery. There's some debate as to whether it actually is; the "No. 1" sign on the label actually means it's the first distillery in Tennessee, but let's not allow the facts to get in the way of a good story!

Business progressed smoothly until 1910 when

Tennessee went dry and the Daniel distillery was forced to switch its operation to St Louis. Jack died a year later from gangrene brought on by icing the door of a safe, leaving the distillery in the hands of his nephew, Lem Motlow.

It wasn't an easy time for whiskey-makers, and when full Prohibition struck Lem turned his hand to horse breeding. However, it's clear that whiskey ran in his veins and after Repeal he started the ball rolling to get Tennessee whiskey being made under the Jack Daniel's name once more. The trouble was that another distillery claimed that it owned the rights to the brand name. The argument went to court and Motlow won.

He had the name, but he didn't have a distillery, as Tennessee was still operating a ban on whiskey production. Undaunted, he took on the legislature and again won his case.

In 1933, distilling started up in Tennessee once more. It wasn't a full victory though. Drinking is still frowned on – there may be millions of gallons maturing at the distillery, but you still can't buy a drink in Lynchburg!

The whiskey itself shows the influence of a high percentage of corn in the mash bill and very little rye. This is a smooth whiskey from the word go. The mellowing vats are also equipped with a woollen blanket just to soften things a little more.

Jack Daniel's Black Label is one of the world's best known brands. Less widely seen are Green Label – a lighter-flavoured, lower-strength version – and the premium variant, Gentleman Jack. This is a barrel selection that has been given a light second mellowing before bottling, which smooths it out even more. It's as smooth as they come.

George Dickel

The German-born George Dickel was a whiskey wholesaler before he bought Tullahoma's Cascade distillery in 1888. It was a shrewd move. Tullahoma was a spa town and the same people who took its waters no doubt sampled its whiskey as well. It was also on the railroad, an important consideration for whiskey-makers who wanted to get their product out onto the American market.

In those days Tullahoma could boast eight distilleries, and by the turn of the century Dickel's Cascade (one of three that took their water from Cascade Hollow) was the largest in the state. It was also the first to advertise its product – at that point still called Cascade – as "Tennessee Whiskey". George, sadly, didn't live to see this happen, because he died in 1894, following a riding accident.

This is our little waterfall down in Coffee County, Tennessee. A hundred years ago, the Freestone Water that comes over this fall made a fine beginning for Mr. Dickel's "drinkin'" whisky. Still does.

George Dickel
Tennessee Sour Mash
Full-bodied enough to start with
light enough to stay with.

The success was short-lived. When Tennessee went dry, Dickel was forced to move and restarted production at the Stitzel-Weller distillery in Louisville, Kentucky – although it retained its own mash bill and charcoal mellowing. Unlike Jack Daniel's it remained in Kentucky after Repeal, though by this time it was owned by Schenley – the same firm that had fought with Lem Motlow over the rights to Jack Daniel's. Schenley restarted production of Dickel, in Frankfort, Kentucky, when distilling started up again after the Second World War. Dickel, the first brand to call itself Tennessee whiskey, was no longer entitled to the appellation. Everything was put right when a new dis-

tillery was built on the original site in Cascade Hollow in 1958.

It's a small-scale operation when compared with Jack Daniel's, but produces a classy whiskey. Like Jack there's only a small percentage of rye in the mash bill, but it's what happens after the new spirit has flowed off the still that makes it different.

For starters, Dickel takes a belt-and-braces approach to mellowing, having woollen blankets at the top and bottom of the charcoal-filled vats. It also ages in single-storey warehouses rather than the common rick-style used elsewhere. This results in a cooler environment, less ingress of the whiskey into the barrels and a more delicate end product.

TENNESSEE WHISKEY

THE CLASSICS

THERE MAY ONLY BE TWO DISTILLERIES LEFT IN TENNESSEE BUT ONE OF THEM PRODUCES THE WORLD'S BEST KNOWN AMERICAN WHISKEY – IRONICALLY MOST PEOPLE THINK OF IT AS A BOURBON! BY FILTERING THE NEW WHISKEY THROUGH VATS OF CHARCOAL TENNESSEE DISTILLERS PRODUCE A STYLE THAT'S GLOSSIER AND SMOOTHER THAN THE BOURBONS FROM KENTUCKY. THESE REMAIN SOME OF THE WORLD'S GREAT SIPPING WHISKEYS THAT HAVE SOMEHOW MANAGED TO COMBINE DOWN-HOME IMAGERY WITH A ROCK 'N' ROLL HEART.

GEORGE DICKEL NO. 8 *40% ABV*

An enigmatic nose that mixes lime blossom and smoky mocha notes. Light fruit on a palate that is both smooth and dry.

GEORGE DICKEL NO. 12 *45% ABV*

A more honeyed nose, with cigar boxes and some of the mintiness you can find in Kentucky. Excellent length.

GEORGE DICKEL SPECIAL BARREL RESERVE *43% ABV*

Oily, and as sweet as a toffee. The most complex of the Dickel brands, it has another dimension of exotic North African fruits and spices. Marvellous.

JACK DANIEL BLACK LABEL *40% ABV*

A sweet, clean nose with hints of liquorice, caramel and smoke. Leaps on to the palate. Ripe, sweet and smooth.

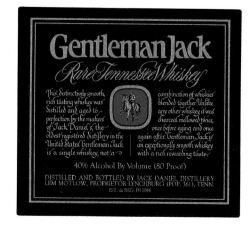

GENTLEMAN JACK *40% ABV*

Not only an altogether sweeter experience than Black Label, but a fruitier one as well. Rich and sultry.

CANADIAN WHISKY

ALTHOUGH FRUIT BRANDIES, APPLEJACK AND RUM WERE THE EARLY DRINKS OF CANADIAN SETTLERS, IT WAS INEVITABLE THAT WHISKY WOULD BECOME THE NATION'S NATIVE SPIRIT. CANADA, AFTER ALL, WAS THE DESTINATION FOR THE MAJORITY OF THE SCOTS FLEEING THE HIGHLANDS IN THE AFTER-MATH OF THE SUPPRESSION OF THE SECOND JACOBITE REBEL-LION IN 1745. IN FACT, AN ORCADIAN EXPLORER HAD ESTABLISHED A SCOTTISH SETTLEMENT IN NOVA SCOTIA IN 1394! EVEN TODAY, GAELIC IS SPOKEN IN CAPE BRETON.

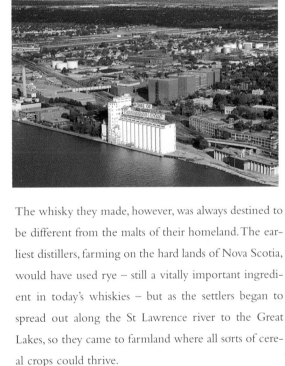

The whisky they made, however, was always destined to be different from the malts of their homeland. The earliest distillers, farming on the hard lands of Nova Scotia, would have used rye – still a vitally important ingredient in today's whiskies – but as the settlers began to spread out along the St Lawrence river to the Great Lakes, so they came to farmland where all sorts of cereal crops could thrive.

Whisky provided a handy extra income for these small farmers who were putting down roots across this vast country, but like every other country the world over, the government soon got wise to the fact and slapped a prohibitively high tax on whisky. The effect was to kill off the small farmer-distiller and concentrate whisky-making in the hands of a few cash-rich producers.

This tax hike came at the same time as continuous stills were being installed – giving these large distillers

LEFT
RIVERFRONT HOME OF
CANADIAN CLUB

OPPOSITE
A NEW WORLD TO CONQUER
FOR SCOTSMEN AND THEIR
NATIVE SPIRIT

LEFT
A CLEAN, EASY-DRINKING
WHISKY

8 5

RIGHT

A STUDY IN STAINLESS STEEL
AT THE CANADIAN MIST
DISTILLERY

an extra incentive to produce from the more efficient, high-volume columns rather than the less efficient pots.

The Canadian government has kept a pretty close watch on the whisky industry. On the positive side, by the end of the nineteenth century, Canadian whisky (it has retained the Scottish spelling) was the most tightly controlled in the world. It could only be made from cereal grains, it had to be distilled in a continuous still and had to spend a minimum of three years in oak.

On the negative side, Canada flirted for a year with Prohibition and then came to its senses, by coincidence at the very same time as the lights were going out on the American whiskey industry. The Great Lakes were filled with boats carrying Canadian (and Scotch) to the thirsty American markets and the industry was saved. It prospered until relatively recently when the downturn in whisky sales forced closures of massive plants across the country.

We have already seen how the whiskies differ from all others. Canadian whisky is the triumph of the blender, just as Irish is the triumph of distillation. For starters there's the huge range of grains that can be used: corn, rye, malted and unmalted barley, and wheat. These can either be distilled individually or in different mash bills, often with their own yeasts. They are all distilled in a column still, but different types of still will be used – linked columns, single columns and doublers as in America, and Coffey stills. The stills can be manipulated to give spirits of different strengths.

These building blocks will then be aged for differing lengths of time, often in different types of wood. Canadian distillers can draw on new, charred and uncharred wood, ex-whisky, ex-sherry, ex-bourbon and ex-rum casks. Sometimes different distillates are aged separately, then blended and aged again, this time in different wood. There's a mass of possibilities.

The whiskies are smooth and soft, but never boring. Instead they are gentle, intriguing and elegant. In today's market, where all the talk is of malt and bourbon, good Canadian whisky remains overlooked. Be bold, try some.

CANADIAN WHISKY

THE CLASSICS

A COMBINATION OF A COUNTRY WHERE ALL MANNER OF GRAINS CAN PROSPER AND GOVERNMENT LEGISLATION HAS GIVEN CANADA A STYLE OF WHISKY UNLIKE ANY OTHER. HERE, THE BLENDER IS KING, CUNNINGLY WEAVING TOGETHER DISTILLATES FROM DIFFERENT GRAINS AGED IN DIFFERENT WOODS INTO BRANDS WHICH HAVE A DISCREET, REASSURING QUALITY. CANADIAN WHISKY HAS LONG BEEN LOOKED DOWN ON INTERNATIONALLY — IT'S TIME FOR A REAPPRAISAL OF WHAT IT HAS TO OFFER.

BLACK VELVET *40% ABV*

Typically soft and smooth. An attractive, rounded feel with the merest hint of rye to pep up the mid-palate.

CANADIAN CLUB CLASSIC

12-year-old (40% ABV)

Rich and elegant with sweet, well rounded, fruity notes. Ripe and smooth on the palate with a tingle of rye, toasty/sawdust wood and long, clean, slightly herbal fruitiness.

CANADIAN CLUB *40% ABV*

A very delicate nose with a hint of smoke. Quite crisp on the nose. Clean, with a little lemony bite from the rye. A clean, easy-drinking (if slightly woody) whisky.

CANADIAN MIST *40% ABV*

Light nose, but a fairly plump palate that gives way to a dry finish. Clean and gentle.

CROWN ROYAL *40% ABV*

A rich, succulent nose, brimful of spicy oak and sweet, almost toffee, fruit. Excellently balanced and elegant, it's a rich, unctuous, mouthfilling whisky.

SEAGRAM V.O. *40% ABV*

An appealing, delicately aromatic nose. The palate is complex, with zesty rye and smooth, mature fruit all on show.

WHISKY TASTING

THE ONLY WAY TO LEARN ABOUT WHISKY IS TO TASTE IT. THE MORE YOU TASTE, THE MORE YOU LEARN, AND SINCE THERE'S ALWAYS SOME NEW BRAND ON THE MARKET AND OTHER BRANDS NEED TO BE RE-CHECKED, YOU NEVER STOP THE LEARNING PROCESS. FEW VOYAGES OF DISCOVERY ARE SO REWARDING.

There is no right or wrong answer in tasting. As you become more experienced, so you'll begin to identify certain aromas and flavours and infer what they mean. In other words, you'll not only be able to say what something smells like, but make a considered guess as to why it smells that way.

How you get there is courtesy of a reference library you've created in your head. While there will normally be some consensus in a panel tasting, each person will have come to their conclusions in his or her own way. Everybody's nose is different, and you may find that you are more sensitive to some aromas than the person next to you. Don't panic if you can't smell the bananas they're going on about – they might not be able to smell the oranges that you can!

Think big. These remarkable spirits have an almost mystical link between aroma and place. Let the pictures pop up in your mind: walking across a Highland moor, standing on a beach in an east-coast breeze, wrapped in a polished leather armchair with a hickory wood fire burning in the grate. Add these pictures to your reference library.

Try comparing different styles. Mix three bourbons, three Irish and three malts. Understand the production methods which yield the different flavours – in time you'll be able to appreciate that the bite in bourbon comes from rye, the fresh crunch in Irish whiskeys is the result of unmalted barley, and how the use of peat and sherry wood affects the aromas and tastes of a malt.

ABOVE
CONJURING GLOWERING MOUNTAINS AND DEEP BLUE POOLS IN YOUR MIND

LEFT
ENJOYING THE *CRAIC*

91

CONNOISSEUR'S TIPS

Where to Taste?

Make sure the environment is right. Don't try to do a serious tasting in a kitchen when there's cooking going on, don't do it in a roomful of flowers, and don't wear perfume or after-shave.

When to Taste?

Most professionals agree that your senses are at their height in the middle of the morning. The anticipation of lunch has got those gastric juices working. While it's great fun to do a spontaneous tasting after a dinner party, don't expect to perform as effectively as you would earlier in the day.

How to Taste?

Use glasses shaped like sherry copitas. They hold a small amount, are easily warmed in the hand and the tapering bowl concentrates and holds in the aromas.

Don't panic! Take your time. It's inevitable that your tasting ability will be better on some days than others, but if you relax you'll soon find you perform consistently. And remember, the more regularly you taste the better you'll get!

STAIRCASE TO HEAVEN

Step 1: Colour

Lift the glass, tip it away from you, and examine the colour. This can suggest a multitude of things, the main ones being the age of the whisky and the type of cask it has been stored in. In general, whisky becomes darker

with age, but the type of wood used also has a part to play. Because bourbon and Tennessee whiskey is aged in new casks they pick up a rich, reddish colour very quickly. These casks may well be sent to Scotland and Ireland to be re-used. Because much of the colour has been sucked out, a 12-year-old malt may be lighter than a four-year-old bourbon. Sherry casks, particularly those made from Spanish oak, will give a massive amount of colour. So be careful about leaping to conclusions!

Step 2: The Look

Swirl the whisky around the glass. You'll notice some clear "tears" clinging to the side. The general rule is the longer the tear the stronger the whisky, and the more persistent they are, the richer the flavour is likely to be. Before you have even smelled the whisky you've already got three clues.

Step 3: The Aroma

Most of the work in whisky tasting is done on the nose. Blenders will never taste – some are even teetotallers! The nose is all you need. If you don't believe me, try pinching your nose very hard while tasting a whisky. Now free your nose and do the same. I'll bet you could taste more the second time.

Step 4: The First Sniff

Don't shove your nose deep into the glass and inhale deeply – you'll kill your sense of smell. Instead give a gentle sniff slightly above the glass, then inside it. Pause and write down what you smell. How intense is the aroma? How complex?

Step 5: The Second Sniff

Add a little water to trigger the more complex flavours. Only use still, bottled water. It's best to start with a tiny drop and, if the aroma is still too fierce, add another. Repeat the same nosing as you did with the unreduced whisky. Don't push it. Relax, let your mind work out what the smells remind you of. Struggling inevitably ends up in big sniffs and numb noses! I prefer to nose all the whiskies before the next step…

Step 6: The Flavour

Take a small sip. What does the whisky "feel" like? Light, oily, hot? Take another sip. This time move the whisky around the mouth. How sweet is it, how savoury? Is it sharp or dry? Every whisky will contain some of these elements, but which ones are dominant? Is the overall impression one of balance? Note what flavours are appearing. Are there some which you picked up on the nose? Then swallow. It's important to do this because the flavour at the back of the palate is a key factor in evaluating a whisky. What is the finish like – hot? dry? sweet? How long does it linger? Take your time. Leave the whiskies in their glasses and return to them. The best will have evolved still further.

Step 7: Conclusions

Don't be shy to share your opinions with others. Whisky is a highly convivial spirit – it's there to stimulate discussion. A friendly disagreement is often the best way to force you to retaste!

OPPOSITE

CHECKING THE LOOK AT WILD TURKEY

BELOW

AFTER THE TASTING … TIME TO RELAX WITH A DRAM

INDEX

PICTURE CREDITS

The publishers would like to thank the following sources for their kind permission to reproduce the pictures in this book:

Austin Nichols and Co.; **Camera Press**/Bryan Alexander; **Campbell distillers Ltd**; **The Chivas and Glenlivet Group**; **Cooley Distillery Plc**; **Corbis UK Ltd**; **E.T Archive**; **Mary Evans Picture Library**; **Glenturret Distillery Ltd**; **Heaven Hill Distilleries. Inc**; **Hulton Getty**; **Images Colour Library**; **Lang Brothers Limited**; **Chris Lobina**; **Macallan-Glenlivet Plc**; **Morrison Bowmore Distillers Ltd**/Eric Thorburn; **Mulcaster**; **Phipps Public Relations Ltd**; **Scotland in Focus**/A.G Johnston; **William Grant & Sons International Ltd**; **United Distillers & Vintners UK**; **Hiram Walker & Sons Ltd**; **Whyte & Mackay Group**

Every effort has been made to acknowledge correctly and contact the source and/or copyright holder of each picture, and Carlton Books Limited apologises for any
unintentional errors or omissions which will be corrected in future editions of this book.

AUTHOR'S ACKNOWLEDGEMENTS

Though sadly there isn't the space to thank everyone who has helped in the writing of this book,
 some have to be singled out: Jim McEwan, Morrison Bowmore; Jim Turle, John Ramsay and Alan Reid, Edrington Group; David Robertson, Macallan; Bill Farrar, Matthew Gloag; Campbell Evans, Scotch Whisky Assoc; Colin Scott, Chivas Bros; Ian Grieve, Jonathan Driver and Nick Morgan, UDV; Stuart Thomson, Bill Lumsden and Kirsty Mellis, Glenmorangie; Mike Nicolson, Lagavulin, Alan Winchester, Aberlour; Euan Mitchell, Springbank; Vanessa Wright, Campbell Distillers & David Hume, Richmond Towers; Jenny Stewart, Seagram; Sheila Reynolds, Allied Domecq; Sandra McLaughlin, JBB; Fergus Hartley, Burn Stewart; Mhairi Adam, Inver House; Jimmy Russell, Wild Turkey; Bill Samuels, Maker's Mark; Gerard White, UDV; Booker Noe, Jim Beam; Jim Rutledge, Four Roses; Max Shapira, Heaven Hill; Bill Creason, Brown-Forman; Neil Mathieson; Nicky Forrest; James Craig-Wood.

To the Scotch Malt Whisky Society for giving me a place to rest my head and whisky to inspire it. To fellow scribes, in particular, Charles MacLean, Susan Keevil, Nigel Huddleston, Jonathan Goodall, Barbara Cormie, Chris Losh, Fiona Sims and Mark Jones.

To Ken Hoskins and Claire Powers in Kentucky for their friendship and teaching me the joys of Manhattans. To my editor, Martin Corteel, who has been astonishingly patient, helpful and ever cheerful. To Justin Downing for his brilliant work on picture research and finally to my beloved wife, Jo, for putting up with me during the writing of yet another book and for being an unfailing, loving source of support.

MIAOW! TOWSER, THE SADLY
MISSED, RECORD-BREAKING
MOUSER OF GLENTURRET